Introduction to Corpus Linguistics

Introduction to Corpus Linguistics

Sandrine Zufferey

iSTE

WILEY

First published 2020 in Great Britain and the United States by ISTE Ltd and John Wiley & Sons, Inc.

ISTE Ltd
27-37 St George's Road
London SW19 4EU
UK

www.iste.co.uk

John Wiley & Sons, Inc.
111 River Street
Hoboken, NJ 07030
USA

www.wiley.com

MIX
Paper from responsible sources
FSC www.fsc.org FSC® C013604

Library of Congress Control Number: 2020938264

British Library Cataloguing-in-Publication Data
A CIP record for this book is available from the British Library
ISBN 978-1-78630-417-9

Contents

Preface

Since the 1990s, linguistics has progressively experienced a fundamental methodological turning point. Following the works of American linguist Noam Chomsky, it changed from the essentially rationalist discipline it had been since the middle of the 20th Century, and gradually (re)opened up the empirical approaches represented by corpus linguistics and experimental linguistics. Over the past decade, this transition has accelerated even more, in such a way that the majority of linguistic works published in international journals currently make use of empirical data. Thus, linguistic corpora have gradually established themselves as fundamental tools for linguists, and their use has spread to other fields in linguistics, including those traditionally favoring a rationalist approach, such as syntax. The development of corpus linguistics has led to the creation of new methods for collecting and analyzing linguistic data, which were made possible thanks to the development of computers and the arrival of the Internet. This new direction in linguistics has encouraged spectacular advances for dealing with the multiple facets of human language in all its complexity from a scientific perspective. Our book intends to introduce such a wealth to readers who are not particularly used to reading linguistics-oriented literature.

In our times, the ability to quantitatively analyze corpus data has become an integral part of the linguist's toolbox. Nevertheless, the use of such data is based on precise theoretical and methodological principles, which require a thorough understanding. This turning point in linguistics implies the need to introduce the new generations of students to the use of these methods which will help them understand the issues underlying their use in scientific literature, to critically assess the results obtained, and to use them in the context of their academic work. Our book is intended as an educational support for students and, in general, for all those wishing to learn the use of corpora in linguistics.

The material introduced in this book does not presuppose prior skills other than basic linguistic knowledge, as well as a minimum command of the most common computer tools, such as spreadsheet software. This book has been designed as study material for teaching corpus linguistics at university initiatory phases, as well as a tool for students wishing to be trained in the use of corpora. Students will be able to work independently thanks the revision questions presented at the end of each chapter, and the detailed answers provided.

As it is an introductory work, this book is necessarily partial and does not deal with all the questions raised by the use of corpora in different linguistic disciplines. It does not cover certain advanced analysis methods which require a high level of computer and statistical skills for data analysis. However, further readings are suggested at the end of each chapter that will enable those who wish to deepen one or other of the aspects presented to go a step beyond.

Finally, this book places a special emphasis on French as an object of study. While it is true that corpus linguistics has imposed itself in an incontestable manner in the English-speaking world and that a significant proportion of French-speaking researchers currently use these methods, the teaching of corpus linguistics still remains marginalized in France. Therefore, this book also aims to highlight the vitality and richness of corpus studies devoted to French, as well as identify the most important resources which have been developed for this language, in the hope of making a contribution to the rise of this discipline for the study of French.

Sandrine ZUFFEREY
June 2020

1

How to Define Corpus Linguistics

This chapter aims to offer the main defining elements of corpus linguistics in order to understand what this field includes. It also aims to lay the theoretical and methodological bases on which the discipline is based. In particular, we will introduce the difference between empirical and rationalist methodologies in linguistics, the important role of computer science for corpus linguistics, the difference between quantitative and qualitative studies, as well as the differences between corpus linguistics and experimental linguistics. In conclusion, we will briefly review the different types of corpora. In the upcoming chapters, this introduction will help us to tackle the research questions that can be answered by means of a corpus study.

1.1. Defining elements

The term *corpus* has a Latin origin and means "body". A *text corpus* literally *embodies* a *set of texts*, a collection of a certain number of texts for study. For example, it is possible to collect a series of newspaper articles and make a corpus of them in order to study the specificities of the journalistic genre. In the field of language teaching, it is also possible to collect texts written by students having different levels, and to build a corpus of these writings in order to study the typical errors that students produce at different learning stages. A methodology using data from the outside world rather than using one's own knowledge of the language is called an **empirical methodology**. Corpus linguistics can be defined as an empirical discipline *par excellence*, since it aims to draw conclusions based on the analysis of external data, rather than on the linguistic knowledge pertaining to researchers.

Working with corpus linguistics therefore implies being in contact with **linguistic data** in the form of texts, and also in the form of recordings, videos or any other sample containing language. Most of the time, these samples are collected in a **computerized format**, which makes it possible to study them more effectively than if they were on paper. Let us imagine, for example, we wish to know how many times and in what passages Flaubert evokes the feeling of love in his novel *Madame Bovary*. If we have a paper version of that book, finding these passages will be a long and tedious task, which will require going through the entire text. However, having a computerized version would make the task much easier. We simply need to look up for the terms *love, in love* or the verb *to love* in its different forms with the search function of the word processor so as to locate the appearances and easily count them. For most of the questions addressed by corpus linguistics, it would be impossible to search through a paper database, and that is why having computerized corpora becomes essential.

The problem of manual tracking and counting of occurrences is all the more acute since corpus linguistics is often based on **large amounts of data** which have not been drawn from a single book, in view of observing the multiple occurrences of a certain linguistic phenomenon and thus apprehending its specificities. For example, let us suppose that we wish to know whether Flaubert talks about love in his work. In this case, focusing solely on *Madame Bovary* would induce a bias, because this novel is not representative of the whole of his work. So, in order to be able to answer this question, it is necessary to go through the entirety of his novels, making the task even more complex to perform manually. Let us now imagine that this time we want to know whether the French authors of the 19th Century all deal with the question of love as much as Flaubert does. In this case, it would be impossible for us to look up the occurrence of terms related to love in all of the novels written by French authors in the 19th Century. In order to avoid this problem, it would be necessary to collect a sample of texts, representative of the works of this period. We will discuss this topic in Chapter 6, which is devoted to the methodological principles underlying the construction of a corpus. For the moment, the important point to bear in mind is that corpus linguistics often resorts to a **quantitative methodology** (see section 1.5) so as to be able to generalize the conclusions observed on the basis of a linguistic sample to the whole of the language, or belonging to a particular language register.

As we will see in the following chapters, corpus linguistics may be of use in all areas of linguistics, for instance in fundamental (see Chapter 2) or applied (see Chapter 3) linguistics. For example, it is crucial in lexicography, since it makes it possible to make an exhaustive inventory of a language's lexicon. It also makes it easy to find examples of uses in different types of sources (literary, journalistic and others), while bringing to light the expressions in which a word is frequently used. In other words, it makes it possible to establish very useful phraseology elements for dictionaries. For example, it is useful to know what the word "knowledge" means, but it is just as important to know that this word is frequently used in phrases such as "acquire knowledge" or "having good knowledge of", etc. Corpus linguistics is a particularly effective method for establishing the frequent contexts in which a word or an expression is used. But corpus linguistics is also used for conducting research in fundamental areas of linguistics such as the study of syntax, since it makes it possible to identify the types of syntactic structures used in different languages. For example, by making a corpus study, it is possible to determine in which textual genres the passive voice is most commonly used. Finally, thanks to the existence of a corpus of oral data, corpus linguistics also makes it possible to answer questions related to phonology and sociolinguistics. For instance, it makes it possible to establish the area of geographical distribution of certain pronunciation traits, such as differentiating the short /a/ form in the French word "*patte*" (paw), from the long /ɑ/ form in the word "*pâte*" (pastry). Answering these different questions requires the use of different types of corpora, as well as having available data regarding their contents. For example, in order to determine the geographical area of diffusion of a certain pronunciation trait, it is necessary to know where each speaker having contributed to the corpus came from. This type of information is called corpus metadata. We will review the main types of existing corpora at the end of this chapter, and discuss the issue of metadata in Chapter 6.

To sum up, in this section, we have defined corpus linguistics as an **empirical** discipline, which **observes and analyzes** quantitative language samples gathered in a **computerized** format. In the following sections, we will discuss in depth the different central points of the definition, indicated in bold, in order to better understand the theoretical and methodological anchoring of corpus linguistics.

1.2. Empiricism versus rationalism in linguistics

Corpus linguistics is an **empirical** discipline, which means that it uses data produced by speakers in order to study language. This methodology is opposed to the **rationalist** method, which functions by looking for answers by relying on one's own linguistic knowledge, rather than looking for it in external data. Let us take an example. In order to determine whether the phrase "When do you think he will prepare which cake?" is grammatically correct or not, the use of empirical methodology would go through large corpora to find whether this syntactic structure is used by English speakers or not.

If sentences following such a syntactic structure never or almost never appear in the corpus, linguists might conclude that this sentence is only rarely used in English. Rationalist methodology, on the contrary, might respond to the same issue by relying on the intuitions of linguists. In this particular case, they might wonder whether they could produce such a sentence or not, whether it seems correct or incorrect depending on their knowledge of the language and might infer a grammaticality judgment from it. Grammaticality judgments are often classified into three types: *correct*, *incorrect* or *marked*, in the event that a sentence may seem possible, but sounds unnatural.

This example illustrates a fundamental difference between empirical and rationalist methodology. While the rationalist methodology leads to the formulation of categorical judgments, the empirical methodology provides a more refined answer to this question, since the observation of corpus data offers a precise indication of frequency, rather than a result in terms of *absence* or *presence*. This is one of the reasons why many linguists currently consider that the empirical methodology better matches a scientific approach (in the sense of confrontation against the facts) than a purely rationalist method for studying language.

Nonetheless, the choice between the use of empirical or rationalist methods is not limited to the field of linguistics. Certain scientific branches such as physics, chemistry, as well as sociology and history are essentially empirical disciplines. In fact, both physicists and historians base their insights on external data, which they collect in the world, in order to build a theory, test it and draw conclusions from it. On the other hand, other disciplines such as mathematics or philosophy are traditionally based on a rationalist approach, since mathematicians and philosophers use their own reasoning to build theories and to draw conclusions, rather than from the collection and observation of external data. Philosophers often resort to thought experiments, but these are not experiments in the empirical sense of the term, because they are based on the reflective abilities of researchers.

1.3. Chomsky's arguments against empiricism in linguistics

Although corpus linguistics has experienced a strong growth over the past 20 years, the empirical grounding of linguistics is not new. Linguists have long used observational data. In the 19th Century, for example, linguists used to work on the comparison of Indo-European languages in an attempt to reconstruct their common origin. Research was based on existing data about the languages spoken in Europe such as German, French and English. Similarly, in the first half of the 20th Century in the United States, the so-called distributionist approach to syntax focused on the study of sentence formation in syntactic structures as they appeared in text corpora, and from there, tried to infer language's general functioning. Around the late 1950s, the use of corpora in linguistics was almost completely interrupted in certain fields such as syntax, following the works of the American linguist Noam Chomsky. In fact, Chomsky defended a strictly rationalist methodological approach to linguistics, and fiercely opposed any use of external data. The objections made by Chomsky against the use of external data in linguistics have been numerous. We will briefly review them, to show in what ways most of them have lost their *raison d'être* in the context of current research.

Chomsky's first objection to the use of corpora, which is also the most fundamental one, is that corpora contain language samples produced by speakers. According to him, linguistics should not focus on the linguistic performance of speakers, but on the competence they have in their mother tongue, something he calls their internal language. Now, here is the problem. When people speak, what they produce (their performance) does not necessarily reflect what they know about their language (their competence). For example, under the effect of stress or fatigue, speakers sometimes produce verbal slip-ups or make language mistakes. From time to time, almost everybody happens to badly conjugate an irregular verb and mistakenly produce the form "*he eated*" instead of "*he ate*". However, if the person who produced this wrong form were recorded, and then asked whether he or she thought he or she had spoken correctly or not, we can almost be sure that he or she would realize his or her mistake and would be able to state the correct form, "*he ate*". Conversely, a speaker could pronounce a word like "*serendipity*" after having heard it from somebody else's lips, but without really knowing its meaning. These examples illustrate the fact that the words speakers "utter" are not always a true reflection of their linguistic competence. In this way, according to Chomsky, the fact of studying corpora places linguists on the wrong track, because they lead them to consider language from the point of view of "production", which merely represents a biased reflection of the rules of language.

According to Chomsky, another problem related to corpus linguistics stems from the fact that corpora are not representative of the language as a whole. He illustrates this problem in an extreme way, by picking the case of an aphasic speaker recorded in a corpus. Linguists analyzing this corpus would draw totally incorrect conclusions about the language in question, since this person does not represent the linguistic competence of a typical speaker. Furthermore, even if we were not to include an atypical speaker, a corpus could never represent more than a tiny language sample when compared to all the oral and written productions in any language. It is for this very same reason that it is impossible to conclude that a word simply does not exist in a language just because it is absent from a corpus. It could simply never have been pronounced in such particular context, while it could exist in other language registers or have been mentioned by other speakers not included in the corpus. This problem is particularly acute in the case of rare linguistic phenomena, such as infrequent words or little used linguistic structures.

This limitation has led to Chomsky's third criticism of corpora, namely the fact that a corpus can never contain the whole of a language and that, therefore, the above-mentioned biases are not solvable. According to him, this problem is all the more serious because even if a corpus were of a very large size and included a representative portion of the language, it would not be fully analyzable by linguists, given the fact that it is impossible to manually analyze the content of billions of sentences.

Chomsky's last two objections have largely become obsolete due to the advances made in computer science. In fact, the size of corpora has increased exponentially over the past 20 years, and corpus analysis tools have also made considerable progress. It has thus become possible to analyze very large amounts of data, which represent a much more accurate mirror of the language than when Chomsky formulated his objections. We will return to this in section 1.4, devoted to the connections between computer science and corpus linguistics. In addition to these technological advances, theoretical and methodological advances have also largely made it possible to eliminate or control the other types of biases mentioned by Chomsky. For example, good practice for building a corpus is to accurately document the type of language it contains. This helps to avoid analyzing the language of a single aphasic subject by mistake, for example, as Chomsky might suggest. It is nonetheless true that a corpus can only show that which it contains, and therefore the absence of evidence that a word or a structure exists in a corpus cannot constitute definitive proof of their absence from the language. Thus, for certain research questions relating to rare or hardly observable phenomena in a corpus, it might be advisable to complement research with another empirical method, namely with the experimental method. As we will see later in this chapter, this method shares the use of a quantitative methodology with corpus linguistics.

In conclusion, we should point out that the rationalist method suggested by Chomsky is also accompanied by biases and limitations which are not negligible and can be corrected by the use of empirical methods. In particular, this method leaves a large space for the subjectivity of linguists while it overestimates the linguistic skills of speakers. Indeed, the use of grammaticality judgments presupposes that all speakers have a definite and consistent intuition regarding all the sentences in their mother tongue. However, such is not the case. If all English speakers agree that a sentence like "Mary dog her walks" is incorrect in English, whereas the sentence "Mary walks her dog" is correct, judgment will not be so unanimous in the case of complex sentences, as the one mentioned above: "When do you think he will prepare which cake?". These divergences become problematic as soon as these judgments are used for building a linguistic theory. What is more, while it is likely that many English speakers would reject a sentence such as "He does be working" for being grammatically incorrect, in certain areas of the English-speaking world (such as Ireland), this sentence would be acceptable. By resorting to many different speakers and including them in reference corpora of speakers coming from different geographical areas, corpus linguistics makes it possible to respond to this problem in a much more satisfactory way.

What is more, in many areas of linguistics such as lexicology, language acquisition and sociolinguistics, the idea of relying on the internal judgments of linguists is simply not conceivable. No one can study children's language by remembering how he or she spoke as a child, or make assumptions about language differences between men and women by imagining how he or she would speak if he/she were a man or a woman. In all these fields, the use of text corpora has been obvious for a long time and corpora use was never interrupted as a result of Chomsky's work. The paradigm shift in recent decades has taken place in areas where it is conceivable to use a purely rationalist methodology, for example syntax.

Finally, it is important to remember that the role of linguistic theory and the intuition of researchers is not absent in most corpora studies. Indeed, a majority of linguists consider corpora studies as a tool, making it possible to validate or invalidate hypotheses on language, formulated in advance, on the basis of scientific literature and their linguistic intuitions. We will see many examples of this approach (empirical validation) throughout this book. This corpus-based research approach is opposed to an approach which considers corpus data as the only point of reference, both in a theoretical and a methodological sense. In this approach, linguists begin their research without an *a priori* and simply let hypotheses emerge from corpus data (this is called a corpus-driven approach). This approach is almost unanimous among linguists working with an empirical methodology. On this point, we agree with Chomsky's metaphorically explained opinion where he states that working with linguistics in this way would be the equivalent for physicists of hoping to discover the physical laws of the universe by looking out of their window. Observing data without a hypothesis often leads to not being able to make sense of data. It is for this reason that the approach that we will adopt in this book corresponds to a corpus-based approach, considering these as available tools for linguists to be able to test their hypotheses.

1.4. Corpus linguistics and computer tools

As we have seen above, corpus linguistics, as performed nowadays, cannot do without computers. Even if works related to corpus linguistics have existed for a long time (such as the indexing of the Bible by theologians or the file-based construction of dictionaries by scholars like Antoine Furetière in French or Samuel Johnson in English), this discipline was only able to properly take off after the arrival of computing.

Corpus linguistics depends on computer science for various reasons. The first one, which we have already mentioned above, is related to the need for computerized texts in order to be able to carry out truly quantitative research. Nevertheless, looking for elements in a corpus, even a computerized one, by using a simple word processing tool is rather inconvenient. Going back to the example of the search for terms related to love in Flaubert, which we discussed earlier, we find that the use of the search function of a typical word processor quickly reaches its limits. First of all, in order to verify that all occurrences found when looking for the verb *to love* correspond to expressions of love as a feeling rather than to modal uses as in the phrase "*I would love it that you kept quiet*", it is necessary to examine each occurrence and thus browse the entire text. Second, to find all the occurrences of the verb *to love*, it is necessary to perform a different search for each verbal form, for example *love*, *loved*, etc. It is for this reason that other computing tools, specifically devoted to corpus linguistics, have been developed.

In particular, **concordancers** are useful for searching all the occurrences of a word, plus their context of use and for displaying the results line by line in a single query. These tools also make it possible to establish the list of words contained in the corpus, together with their frequency, and to generate a list of keywords matching the content of a corpus. In the case of corpora containing texts as well as their translation, certain tools called aligners make it possible to align the content of the corpus sentence by sentence. That being done, bilingual concordancers search directly for the occurrences of a word in one of the two languages of the corpus, and simultaneously extract the matching sentence in the other language. We will learn how to use these tools in Chapter 5, which is devoted to the presentation of the main French corpora, as well as the tools for analyzing them.

Then, in Chapter 7, we will also see that in order to answer certain research questions, it is necessary to annotate the content of a corpus. For example, let us imagine that we wish to study the different contexts in which we can use the causal adverb *since*. If we only look up the word *since* in the corpus, we will also find occurrences which do not correspond to the use of this word as a causal adverb, but to its use as a preposition, for example in "I haven't seen Mary since Christmas". So, to be able to correctly look up the uses of *since* we are interested in, we should only keep those which are adverbs and exclude prepositions. This search can be greatly simplified if the corpus has been annotated by determining, for each word, its grammatical category. This operation, called part-of-speech tagging, can be performed automatically by certain software.

Another problem might arise if we decide to study the use of relative phrases such as "the girl who is intelligent" or "the violin which was left on the bus". For this study, a good starting point would be to look for relative pronouns such as *who* or *which* in order to find occurrences of relative sentences in the corpus. The problem is that these pronouns are also used in interrogative sentences such as "Who do you prefer?" or "Which hat is yours?" In this case, looking for the grammatical category of the word will not solve the problem, because they are both pronouns. In order to find only the occurrences of *who* and *which* as relative pronouns, we should use a corpus in which the syntactic structure of each sentence has been analyzed in such a way that we can assign a grammatical function to each word and group them into syntactic constituents. Tools for analyzing the syntactic structure of sentences have also been developed in the context of works for automatic language processing. These automatic analyses still require human checks so as to avoid any form of error, but their performance is continually improving. The arrival of these tools has greatly accelerated research in corpus linguistics. We will discuss this issue in Chapter 7, which is devoted to annotations.

But corpus linguistics was not only developed thanks to the creation of such tools. Above all, it is the general development of computers and the digital revolution which have made the greatest advances possible. In fact, the increase in the computing power of machines – as well as in their memory – has made it possible to build ever larger corpora. Until the 1980s, a corpus of a million words was considered to be a very large corpus. For instance, the first reference corpora (such as the *Brown* corpus developed for American English in the early 1960s) were about this size. At the same time, the arrival of cassette recorders to the market enabled the first creations of oral corpora containing an exact transcription of spoken speech, rather than a synthesis taken in shorthand.

The marketing of scanners in the 1980s later made it possible to digitize a significant amount of data and corpora began to reach larger sizes, up to 20 million words. Then, with the democratization of computer use, the amount of digitally disseminated texts greatly accelerated the growth of corpora.

Finally, since the beginning of 21st Century, the wide dissemination of documents online via the Internet has given another dimension to the size of corpora available to researchers. At present, the *Google Books* corpus, for example, contains more than 500 billion words, which represents approximately 4% of all the published books of all time (Michel *et al.* 2011). We will discuss the possible uses of such a corpus in the following chapters. In Chapter 6, we will also see that the Internet potentially offers an exceptional data resource for corpus linguistics, but that Internet research cannot be used without an additional processing step if we are to grant data quality.

1.5. Quantitative versus qualitative methods

We have seen that computers help us to work on very large corpora and automatically count word occurrences, find keywords, etc. The need to use a large amount of data and the desire to quantify the presence of linguistic elements in a corpus corresponds to a quantitative research methodology. This methodology involves observing or manipulating variables, as well as the use of statistical tests. The main objective is to test a limited number of variables, in a highly controlled environment whenever possible and on a language sample that can be representative of the phenomenon studied. This can later make it possible to generalize the results obtained to the whole language or to a part of the target language (e.g. journalistic language). These methods nonetheless imply a certain form of reductionism and a simplification of reality. Ultimately, the addition of studies with well-defined and properly controlled variables may provide a global and realistic picture of a phenomenon.

Let us take an example. Suppose we want to test the hypothesis that women talk more about their feelings than men. To test this hypothesis by means of a corpus study, we should first make sure that we are comparing records of men and women produced in the same context, for example, in the context of friendly discussions around a topic, or a face-to-face interview with a researcher. We will also need to make sure that the corpus collected in this way includes approximately the same speaking time or the same number of words pronounced by men and women. This control over the linguistic context and the duration of interactions helps us to ensure that men and women have had fairly equal motives to pronounce words related to emotions/feelings, and as many chances of doing so. Second, we would have to choose a list of words to search within the corpus, representative of the vocabulary related to emotions, for example verbs such as *to annoy*, adjectives like *furious* or nouns like *anger*. Then, by comparing the number of times these words have been produced by the two groups and by validating the significance of the differences observed between the groups through statistical tests, we would be able to provide an answer to the research question. In this study, we have sought to reduce the number of confounding variables by controlling the context of production of the statements, as well as by limiting the word choice in the examined vocabulary. It is precisely this limited and reductionist aspect that the opponents to quantitative methods criticize, thinking that the constructed and unnatural context in which structured interviews take place does not reflect the richness of natural and spontaneous exchanges between speakers.

The other major methodological paradigm includes so-called qualitative studies. The main objective of these studies is holistic: they aim to study a phenomenon understanding it as a whole, as detailed and as thoroughly as possible, but in a small number of people. Due to their nature, qualitative studies are interpretative. In linguistics, research paradigms involving a qualitative methodology typically resort to the administration of questionnaires with open questions, interviews, observations or introspective techniques, such as think-aloud protocols. For example, in order to study the differences in the way of expressing emotions between men and women, a qualitative methodology could involve asking a reduced number of speakers, for example three men and three women, to describe the way in which they express their emotions, either by talking freely with the experimenter or by talking to each other. The analysis would then require an in-depth study of some of the examples found interesting during the discussion.

One of the main criticisms aimed at qualitative methods is that they are very subjective in nature, insofar as they are largely based on the interpretations made by linguists and the subjective impressions of a few speakers. Thus, the specific cases they describe cannot often be generalized to a population, which, by the way, is not the aim pursued by such studies. Rather than the generalization of results, these studies are based on the possibility of making a transfer from a particular situation so as to understand another one with which it shares common traits. For example, an in-depth case study on the difficulties of expressing emotions in an aphasic patient may help to highlight similar difficulties existing in other patients with the same disorder.

To summarize, each of the two methodological paradigms introduced in this section has both advantages and disadvantages. Quantitative methods enable the generalization of results to the whole of a population, whereas qualitative methods offer a more detailed and nuanced panorama of a real case. Recently, the complementarity between these approaches has started to be broadly accepted in research and many studies are crossing the two types of methodologies, in order to benefit from their advantages and limit their disadvantages.

For example, if we want to know whether learners of French as a foreign language at an advanced level are able to use collocations as native speakers do (collocations such as "*prendre une décision*" – to make a decision – or "*pleuvoir à verse*" – to pour with rain), we can search for occurrences of these expressions in text corpora produced by learners and compare the number of times these expressions appear – and their frequency – in a corpus of similar textual productions made by native speakers. By comparing these frequencies through statistical tests, we will know whether learners actually use these expressions as often as native speakers do, or not. Even if we find a difference between the two groups, something which this study will not tell us is why learners do not use these expressions as often as native speakers do or which expressions they use instead. To find out, we can complete this study with a qualitative analysis, by observing, for example, which words often accompany the occurrences of the noun *décision* in French, which are not the verb *prendre*. If we observe that several times the verb used is *faire* (make), rather than *prendre* (take), a decision in English-speaking learners, but not in German-speaking learners, we will conclude that these errors could come from a problem of transfer from their mother tongue and, more specifically, from the expression *to make a decision* in English.

In summary, a corpus can be analyzed using a quantitative or qualitative methodology. While we acknowledge the use and importance of combining these two approaches, in the rest of the book we will focus on the quantitative approach to corpus linguistics, which poses its own theoretical and methodological challenges.

1.6. Differences between corpus linguistics and experimental linguistics

Corpus linguistics and experimental linguistics share very important methodological properties, since both are empirical in nature and both generally involve a quantitative rather than a qualitative approach. However, these two types of approaches differ in one very important point. On the one hand, corpus linguistics focuses on **data observation** as found in collections of texts, recordings, etc. On the other hand, experimental linguistics points to the **manipulation** of one or more variables in order to study their effect on other variables.

Let us imagine once again that we are interested in the types of language errors produced by learners of French. By means of a corpus study, we will be able to identify all the types of errors produced and then quantify each of them: for example, 30 spelling mistakes, 12 lexicon errors, 20 syntax mistakes, etc., made every 100 words. Then, by applying statistical tests, we will be able to determine whether one of the error categories is significantly more frequent than the others. We will also be able to compare the number of errors produced in each category by students of different levels and, thanks to statistical tests, determine whether students make significant progress faster in certain categories than in others. In contrast, what a corpus study will not help you to do is establish with certainty the factors influencing the number of errors. The corpus only shows you the result of the speakers' production, but not what led to these results. In order to determine the factors that lead learners to make mistakes or not, we will need to resort to experimental methodology.

When we conduct an experiment, the goal is to manipulate the possible causes and then to observe their effects. Going back to our example research question, we may wonder what makes some students produce more errors than others, and in certain contexts, what makes the same student produce more errors than in other contexts. As regards the difference between students, we may think that one possible cause is the level of general intelligence of each student, the assumption being that overall smarter students should produce fewer errors than less intelligent students. The level of intelligence thus constitutes the cause that we will manipulate in order to observe its effect on the number of errors produced. In order to measure the effect of the intelligence variable, we will first need to measure the students' intelligence, for example by means of an IQ test. We will then use the result of this test to determine whether the students who have a higher IQ are also the ones who make the fewest language errors.

In the case of the second research question, which seeks to determine why the same student makes more mistakes in certain contexts, we may assume that stress promotes the production of errors. In order to test this hypothesis, we will have to conduct an experiment in which half of the students are placed in a stressful situation such as an examination context or, for instance, a test with a limited amount of time to complete the task, whereas the other half of the students are placed in a low-stress situation, for example, without any time constraint, performing a task which does not involve marked assessment, etc. Then, we will compare the number of errors in the two groups so as to determine, by means of a statistical test, whether the students under a stressful situation make significantly more errors than the other students, or not. In the two examples of studies that we have just discussed, the approach is the same: to identify a possible cause and to assess its effect through experimental manipulation. Conversely, a corpus study focuses on linguistic productions without manipulating the data before collecting them.

The study of linguistic productions in a corpus and the manipulation of experimental variables both have their advantages and disadvantages. On the one hand, corpus linguistics has the advantage of favoring the observation of natural data, that is, those which are not influenced by an experimental context. A corpus of journalistic texts includes real productions by journalists, which are not produced for the purpose of being observed. Likewise, a text produced by a learner is also natural, insofar as it is produced in its usual conditions, without there having been any particular manipulation. In addition, the use of corpora favors the observation of a very large amount of linguistic data, whereas experiments are based on a limited number of linguistic items for the task to remain feasible for participants, who would not be able to read thousands of sentences at a laboratory, for example. Finally, once a corpus has been created, it can be used for numerous research questions without requiring any additional time or financial costs. On the other hand, experiments require significant time resources as well as the usual obligation of having to financially compensate participants for their cooperation.

Experimental studies also have definite advantages over corpus studies. The first advantage, mentioned above, is that experiments allow us to test the existence of a causal relationship between two variables, such as the fact of being stressed and producing more errors. Corpus studies do not make it possible to draw this type of conclusion. Second, while an experimental paradigm can be developed to test almost any kind of phenomenon, there are some rare linguistic phenomena which may be absent or too little represented in a corpus to be examined in this way. For example, if we want to decide whether learners are fluent in French idioms such as "*mettre le feu aux poudres*" (to stir up a hornet's nest) or "*avoir un poil dans la main*" (to be extremely lazy) through a corpus study, we will have to look for them in a corpus of learners' productions. Now, it is quite possible that these expressions are never found there, but this does not necessarily mean that the learners do not know how to use them. It only means that they did not have an opportunity to produce them in the corpus. Using experimental methodology, we will be able to test whether learners have mastered these expressions. For instance, we can encourage them to read the expressions and then ask them to choose, from among several definitions, the one corresponding to their meaning. Finally, experimental linguistics makes it possible to study the linguistic competence of speakers, through different language comprehension tasks which can be more or less explicit or implicit, such as the conscious evaluation of sentences, their intuitive reading, etc. Corpora can only reflect the linguistic productions of speakers.

To conclude, corpus studies and experimental studies can often be used in a complementary way, and, when put together, they represent powerful tools for answering a good number of research questions.

1.7. Different types of corpora

As we will see in the following chapters, corpora represent linguistic samples of a very varied nature, and it is precisely this variety that makes it possible to answer diverse research questions in all fields of linguistics. In this last section, we will introduce a first classification of the types of existing corpora, in order to be able to refer back to it in the following chapters.

The first distinction we can make among all the existing corpora is the one that classifies them into a **sample corpus** and a **monitor corpus**. Sample corpora are those in which data have been collected once and for all, and which no longer evolve thereafter. For this reason, they are also known as closed corpora in the specialized literature. The advantage of these corpora is that they have been designed to contain a set of texts representative of the language, or a part of the language to be studied, with a balanced representation of the different text genres, for example. Thus, these corpora make it possible to draw conclusions which can be generalized. On the other hand, their main defect is that they age quickly and do not follow changes in the language. Therefore, sample corpora need to be recollected at regular intervals.

On the other hand, monitor corpora are never finished and constantly continue to integrate new elements, which is why they are described as open corpora in the literature. A typical example of this type of data is the corpus that contains newspaper archives or parliamentary debates. Every year, the number of available data increases. It is for this reason that it is difficult to maintain a perfect balance between the different parts of these corpora, whose representativeness cannot be fully guaranteed. We will return to the problem of representativeness in Chapter 6. On the other hand, these corpora remain up to date. In cases where they comprise a period of a few decades, they make it possible to observe the appearance of certain changes in language.

The second major distinction to be made among existing corpora differentiates **general language** corpora from **specialized language** corpora. General language corpora aim to offer a panorama of the whole of a language at a given time. It is evidently impossible to collect a sample of the whole language, but in the same way that a general language dictionary aims to describe the common lexicon of a language, the general corpus seeks to offer a global image, including the main textual genres found in language. These corpora are really valuable when it comes to studying a language as a whole, but they cannot offer precise answers on linguistic phenomena present in certain specific communication means, such as mobile texting, social media, medical reports, etc.

In order to study one of these areas specifically, it is preferable to resort to a specialized corpus. In fact, there are corpora especially devoted to texting, social media, etc. In addition, general corpora include productions by adults who are native speakers of the language represented. Other corpora specialize in representing other population categories, regardless of whether they are monolingual children in the process of acquiring their mother tongue, bilingual children, foreign-language learners, or even children with neuro-developmental disorders influencing language acquisition, such as autism and specific language impairment. Finally, by default, a general corpus includes examples of the variety considered as a language standard, or one of its main varieties. In French, it generally refers to the French language from France and, more precisely, from the Parisian region. In English, general corpora can refer to the English language from the UK or to American English. Conversely, some corpora specialize in the productions of speakers of a certain language variety, such as French from French-speaking Switzerland, Belgium, Canada, etc.

General or specialized language corpora can contain either **written language** or **spoken language** samples. For a long time, written language corpora were the norm, but analysis of the spoken language has developed broadly since the 2000s. Corpora of spoken language are typically of smaller size than written language ones, since they require manual transcription. As a matter of fact, it is easy to record voices, but what is difficult is to carry out searches directly on an audio file. At the same time, speech recognition software does not always fully allow reliable automatic transcriptions. It is for this reason that the oral data must be transcribed manually, which often limits the size of the spoken corpora. More recently, audio-visual recording corpora (also called "multimodal" corpora) have been created, in order to facilitate, for instance, the study of gestures and facial expressions as well as their role in communication. These corpora still pose many codification and interpretation challenges. Finally, let us point out that video corpora are also used for the study of sign language.

Another distinction that can be made regarding the types of existing corpora relates to the type of processing carried out on the linguistic data of the corpus. On the one hand, **raw corpora** contain nothing but language samples. This scenario represents the majority of the French corpora. On the other hand, some **annotated corpora** contain specific linguistic information, apart from the language samples. The most common type of annotation is the assignment of a grammatical category to each word in the corpus, as we have already mentioned. More rarely, certain corpora contain a syntactic analysis of all of their sentences, as well as other types of information, such as an annotation of the discourse relations (cause, condition, etc.) which interconnect the sentences within the text corpora. Finally, certain corpora, which have been transcribed with the aim of studying phonological phenomena, may end up being transcribed using the International Phonetic Alphabet.

So far, all the types of corpora we have considered are **monolingual**. Another distinction that we can make is to differentiate these corpora from **multilingual** corpora. There are two types of multilingual corpora. On the one hand, we have **comparable** corpora, which contain similar samples produced by native speakers in two or more languages. For example, it is possible to build a comparable corpus of parliamentary debates in France and the UK. Such a corpus would make it possible to compare the ways of speaking in a similar context in two languages and two different cultures. On the other hand, so-called **parallel** corpora contain texts produced in one language and their translation into one or more other languages. These corpora make it possible to study the linguistic correspondences between languages, as well as the linguistic phenomena linked to the translation process. Parallel corpora can also be annotated with exact matches between sentences. This process is called alignment and gives rise to so-called aligned corpora.

Finally, many corpora are drawn from contemporary written or spoken data. However, there are archives that make it possible to study the history of a language, going back to ancient French, for example. Contemporary corpora are used for studying language in a **synchronic** way, that is, at a given moment during its evolution, whereas historical corpora make it possible to carry out studies from a **diachronic** point of view, that is, on the evolution of language.

1.8. Conclusion

In this chapter, we have defined corpus linguistics as an empirical discipline, that is, based on the observation of real data. We have also seen that corpus linguistics often resorts to a quantitative methodology, studying a large sample of data which is representative of the phenomenon studied, with the aim of generalizing the observations to the whole of the language or to a language's register. We have shown that the main difference between corpus linguistics and experimental linguistics is the way in which empirical data are collected. In the case of corpus linguistics, data are collected in a natural context and then observed, whereas in the case of experimental linguistics, one or more causes are manipulated within a controlled context in order to observe their effects. Finally, we have seen that corpora can be very diverse in nature, depending on whether they are made up once and for all or incremental, general or specialized, annotated or not, monolingual or multilingual, synchronous or diachronic.

1.9. Revision questions and answer key

1.9.1. *Questions*

1) Which of the following disciplines traditionally involves a rationalist methodology, and which disciplines are based on an empirical methodology? Can we think of any situation in which a discipline of a rather empirical nature could have recourse to a rationalist methodology and vice versa?

chemistry – ethics – medicine – law – anthropology

2) Among Chomsky's objections to corpus linguistics, which of them can also be applied to the experimental methodology?

3) In the research projects mentioned below, which one seems to use corpora as a methodological tool (*corpus-based*) and which seems to use corpora as a theoretical tool (*corpus-driven*)?

a) Search in a corpus for all passive voice sentences in order to formulate the rules governing the use of this construction in French.

b) Search in a corpus for all passive voice sentences in order to determine whether they are used more with state verbs than with activity verbs.

4) Why have computing tools especially devoted to corpus linguistics been developed? What are their main functions?

5) Look for an example of a quantitative study and another qualitative study that could be done so as to determine the most common types of spelling mistakes made by children. Which would be the specific contributions of each of these studies?

6) How could we use a corpus and carry out an experiment to study the question of the different types of spelling errors in a complementary way?

7) What type of corpus should be used to work with the research questions stated below?

a) Study of the pronunciation of vowels in French-speaking Switzerland.

b) Study of the evolution of word construction using the prefix *hyper-* in French.

c) Study of possible translations of idioms from French into English.

1.9.2. *Answer key*

1) First of all, let us recall that the rationalist methodology interrogates the knowledge of the researcher by means of introspection and reasoning, whereas the empirical methodology looks for answers by observing or experimenting on data that is external to the researcher. Chemistry is typically an empirical science, which makes extensive use of experimentation and observation. Ethics is a philosophical discipline that involves reflections on moral questions. These reflections are, by nature, introspective and involve a rationalist methodology. Law is a science that studies the rules and laws that govern social relationships. Many aspects of the law involve the interpretation of existing rules or the creation of rules based on reasoning and common sense. Thus, introspection plays a big role. That being said, in certain cases, law also deals with external data. For example, a search can be performed throughout previous decisions (case law), in order to find a similar case that could apply to a certain situation. The role of case law is very different in different legal systems. In English-speaking countries, which apply the *common law*, previous cases play a fundamental role, because they become binding rules for solving the following cases. We can therefore say that in these countries, the part of empiricism when applying the law is also very important. Anthropology is a discipline that studies humanity in its various aspects (physiological, social and cultural). This discipline places great importance on the observation of data. Despite the fact that we can generally classify a branch as being rather empirical or rationalist in nature, we should bear in mind that these two methodologies are often present in varying degrees. For instance, we have already discussed the case of law, where not only an introspective element is involved, but also the use of external data in the form of case law. We can also imagine other situations of interaction between methodologies. For example, we have classified ethics as a rationalist discipline. Nevertheless, ethics was also built on the basis of empirical material. In the field of medicine in particular, medical ethics is based on the facts observed in practice.

2) Chomsky notably criticized corpus linguistics for offering only a partial vision of language, insofar as a corpus includes the productions of a limited number of speakers, at a given situation. This same observation also applies to the experimental methodology, which tests a small number of speakers along a limited number of linguistic stimuli. The main response to

such criticism is that these areas are based on the use of quantitative methods (namely inferential statistics), which make it possible to draw conclusions from a sample and to extrapolate them to an entire population. The criticism of the potentially problematic choice of subjects who could be aphasic and not represent the normal use of language also applies to experimental methodology. In theory, though, such subjects could also be recruited for an experiment by mistake. That being said, good practices in corpus linguistics and experimental linguistics require obtaining information about participants beforehand, which can eventually eliminate this type of bias. Typically, researchers verify that the people who contribute to a French corpus are native French speakers. Likewise, they test the language skills of speakers before considering them by default as French-speaking, bilingual, etc.

3) a) This type of research is *corpus-driven*, because the starting point is not hypotheses which have to be verified throughout the corpus. The starting point for research is the corpus itself, in order to be able to infer usage rules from its content.

b) On the other hand, this type of research is *corpus-based*, because it starts from a hypothesis (e.g. "passive sentences tend to be used more frequently with state verbs"), and seeks to verify it in the corpus, which, in that way, only works as an analysis tool.

4) These tools have been developed for simplifying searches within a corpus. Otherwise, it would be very inconvenient to use the standard tools that are present in a word processor, for example. In particular, concordancers make it very easy to extract all the occurrences of a word or an expression with its left and right context, as well as to determine its main collocations. These tools also help us create a list of all the words in the corpus, sorted by frequency. While one corpus can be compared to another reference corpus, these tools also make it possible to extract a list of keywords that are specific to the corpus studied. In the field of multilingual corpora, aligners make it possible to align parallel corpora sentence by sentence, and then to extract a sentence and its translation by means of a bilingual concordancer.

5) A quantitative study on this question could focus on the creation of categories for classifying spelling mistakes, for example, agreement errors, redoubling of consonants, dumb letters, etc., and then counting all the occurrences of errors belonging to each category. By applying a statistical test, this study would then make it possible to know whether students tend to

make certain types of mistakes more often (e.g. grammatical errors) rather than other mistakes (e.g. lexicon errors). A qualitative study on this same question would identify some examples of spelling mistakes and analyze in detail the contexts in which they occur, for example the grammatical category of the words concerned, whether they are rare or frequent words, occurring in a long or a short sentence, what type of phonemes is poorly transcribed, etc. This study would make it possible to identify linguistic contexts that tend to be conducive to spelling mistakes.

6) The results of the quantitative corpus study summarized above, namely the quantification of the different types of spelling mistakes, could be considered as a kick off for an experimental study. For example, the corpus study could help identify one type of common error, and one type of rare mistake. An experiment could then help to determine whether being in a stressful situation or not has a different impact on the two types of error.

7) a) In order to study a phonological phenomenon like this, a spoken corpus is essential. This corpus should be specific to the population of French-speaking Switzerland. A large type of corpus comprising a large number of different speakers would be desirable. Finally, this corpus should contain a synchronic type of data, corresponding to the current pronunciation, rather than to its diachronic evolution.

b) In order to study the evolution of a language, a diachronic corpus is essential. This constraint implies the use of written data, since oral data only go back to the middle of the 20th Century. Finally, the chosen corpus should include productions made by adult native speakers.

c) In order to study translation, a parallel corpus is required. This corpus should contain original texts in French and their translation in English. It should be a synchronic corpus, corresponding to current uses of the language.

1.10. Further reading

For a discussion regarding the main defining elements of corpus linguistics, the works by Habert *et al.* (1997) are an excellent introduction in French, even if tools and methods have evolved considerably since they were published. On Chomsky's objections to corpus linguistics, refer to the book by McEnery and Wilson (2001, Chapter 1). A more detailed discussion with possible responses of corpus linguistics to these objections can be found in Aarts (2001). As regards the role of computer science in the evolution of corpus linguistics, as well as a typology of corpora, see Tognini-Bonelli (2010). On the differences between qualitative and quantitative methods in linguistics and the ways to combine them (refer to Litosseliti 2018). For the basic principles of the experimental methodology see Gillioz and Zufferey (forthcoming).

2

How to Use Corpora in Theoretical Linguistics

After having introduced the theoretical and methodological foundations of corpus linguistics in Chapter 1, in this chapter and in the following one, we will illustrate the multiple uses of corpora in linguistics. In this chapter, we will focus on the fields belonging to theoretical linguistics, namely phonetics and phonology, morphology, syntax, lexicon, discourse analysis, pragmatics, sociolinguistics and the study of linguistic evolution from a diachronic viewpoint. For each study area, we will start with a general presentation of the possibilities offered by the use of corpora, and will be illustrating these possibilities with some examples performed on French corpora. For the moment, we will limit our presentation to monolingual corpora. Multilingual corpora and their uses will be discussed separately in Chapter 4.

2.1. Phonetics and phonology

Phonetics studies sounds as acoustic units, whereas phonology studies how these sounds are used to form meaningful units such as words in different languages. These branches evidently require one to study samples of spoken language that reflect the way in which sounds are produced by speakers. To do this, different empirical methods can be used. The simplest method is to ask a group of speakers to read aloud a list of words or texts, which have been specifically designed to include multiple occurrences of specific phonemes in accordance with the aims of the study (regional and social variations, etc.). In addition to the data provided by readings, the more natural data contained in spoken corpora also represent valuable tools, since they provide different and richer data compared to words or texts read aloud. These data are clearly more representative of spontaneous language use. Indeed, speakers do not behave in the same way when they read or when they chatter spontaneously, and these differences are reflected in the way they pronounce words, as well as in their speed of articulation, as we will see below.

The corpora containing spontaneous interactions in a conversational context have also made it possible to study the various ways in which speakers attune their speech to their conversational partner, as well as the study of prosodic phenomena in various communicative contexts. These studies have shown that, in the context of spontaneous speech, speakers tend to stress those words that are not foreseeable in conversation and to choose a prosodic contour reflecting the syntactic organization of their sentences (Ladd 2008). The diversity of speakers represented in corpus data has also made it possible to study the differences in sound production among different categories of speakers: differences between men and women, the young and the old, as well as pronunciation differences between different regional varieties of the same language.

Finally, the use of corpora makes it possible to perform a quantitative analysis of word frequency, as well as its influence on their pronunciation. Frequency studies have shown that words that are more frequently used have a greater impact on consonant lenition phenomena. For example, the pronunciation of [t] evolves towards [d] and reduces vowels, which are pronounced weakly in certain linguistic contexts (Bybee 2001).

These phenomena are important for understanding how phonological changes occur in a language over time. In summary, studying phonology using corpus data makes it possible to bring to light the interfaces between syntax and speech. These are not so evident when we work with data drawn from reading words or texts (these are less extensive and not so spontaneous).

The first study that we will introduce in this section concerns the question of speech articulation rhythm in different French varieties. Schwab and Avanzi (2015) tested the popular belief that certain varieties of French, notably French from French-speaking Switzerland, might be spoken more slowly than other varieties. In order to test this hypothesis, the authors retrieved three-minute speech sequences from the *Phonology of Contemporary French* database, which were produced by people from different French-speaking regions (eight speakers per region):

– Paris and Lyon (France);

– Tournai and Liège (Belgium);

– Neuchâtel, Nyon and Geneva (Switzerland).

Each person was recorded both during a reading task and throughout a conversation. This study thus made it possible to compare two kinds of speech. Before proceeding with the analysis, the data were transcribed and the texts were aligned with the sound signal. For each speech sample, the authors counted the number of syllables spoken between two pauses. The articulation speed was then calculated in milliseconds per syllable for each segment between two pauses. The data obtained were analyzed using a statistical model which made it possible to test the influence of several variables (the causes we mentioned in Chapter 1; see also Chapter 8 for a more detailed explanation of this notion) on other variables (the observed effect). In this study, the observed effect was the rhythm of speech articulation. The possible causes tested in the model included social variables such as age, geographic origin of the speakers and their gender, as well as speech style (reading or conversation). The results, which we will only partially report here, indicate that the geographic origin of the speakers does have an influence on articulation speed. More specifically, Swiss speakers have a slower speech articulation rhythm than French speakers (particularly Parisian speakers). Belgian speakers also speak more quickly than Swiss speakers, especially those from Neuchâtel and Nyon. This study tends to confirm the idea that Swiss speakers have a slower articulation than other French speakers. Furthermore, speech style may also have an effect on articulation speed, syllable duration being shorter in conversations than in reading. This result confirms the importance of studying not only reading texts but also, more importantly, spontaneous spoken speech excerpts in order to study a phenomenon such as articulation speed.

The second case study that we will discuss in order to illustrate the role of corpora in phonology concerns the phenomenon of *liaisons*. In French, many *liaisons* are considered optional, insofar as speakers can choose whether to make them or not. For example, in (1), it is possible to either pronounce the latent-word final consonant [z] as connected to the final -s of the verb *allons* (let's go) or not.

(1) *Nous allons au cinéma.*

(We are going to the movies.)

Several studies have used corpus data to study the factors that lead speakers to pronounce the *liaison* or not. For example, Meinschaeffer *et al.* (2015) tested the role of linguistic and sociological factors. Their study uses part of a corpus of conversations drawn from Romance languages (*C-Oral-Rom*). In order to identify the words where *liaisons* could be pronounced or not, the authors began by choosing a small number of speakers in the corpus for whom they manually identified all the occurrences where a *liaison* was possible, finding a total of 1,219 locations in the transcripts. Then, by listening to the sound corresponding to each place spotted in the transcription, the authors annotated whether the *liaison* was pronounced or not. For each case, they also took down notes of the speaker's sociological information (age, sex, level of education), the type of speech context (dialogue or monologue) and other information related to the linguistic context (number of syllables in the word, grammatical category, etc.). The authors also wrote down the number of cases where the *liaison* was made in a variable manner (544 cases in total) and paid attention to certain linguistic characteristics of the words involved.

Results showed that optional *liaisons* were mostly found in consonants [t], [z] and [n], almost always on monosyllabic words, and most often on functional words rather than on lexical words. Within the same category, however, large differences could be observed between words, which indicated that language use also plays a key role. No significant difference was observed between the speech contexts (dialogues or monologues). From a sociolinguistic point of view, results indicate that women make significantly more *liaisons* than men, that younger speakers (below 40 years) tend to make more *liaisons* and that older speakers and less educated speakers (who do not have a university degree) tend to make more *liaisons* than more educated speakers. However, the propensity to make *liaisons* or not is also highly variable between speakers belonging to the same sociological category. As the difference between men and women was really significant, the authors also tested the differences between other variables separately for the two genders. These analyses showed that the difference in educational level tends to be significant in women, but not in men. Conversely, the effect of age is significant in men, but not in women. Since this study was carried out on a limited number of participants, it should be replicated before definitive conclusions on the role of these variables can be drawn, especially considering that large variations between participants have been observed. What we should bear in mind is that this study illustrates the way in which linguistic and social variables – which can be studied by means of a corpus – are involved in the production of a phonological phenomenon such as *liaisons*.

2.2. Morphology

In order to study the rules for assembling morphemes into words, as well as the productivity of different morphemes (the number of words that a rule makes it possible to create), morphology needs to rely on external data, since frequency data cannot be inferred through introspection. Therefore, morphologists have long resorted to word lists. At present, these lists can be retrieved automatically from computerized dictionaries. Some morphologists also consider these lists as a form of corpus, because they make it possible to gather large amounts of data. Working with word lists can certainly be considered as an empirical method. Nevertheless, these data do not necessarily represent a corpus *stricto sensu*, since they do not contain extracts of language produced naturally. For some research questions, these lists can be very useful, whereas in other cases, it is necessary to resort to a real corpus containing linguistic productions in their context of use.

The usefulness of word lists has been clearly illustrated by a study by Lyster (2006). This author used a list of nouns drawn from the computerized version of the *Robert Junior Illustré* dictionary so as to determine whether there are certain linguistic regularities that make it possible to predict the grammatical gender of nouns in French. Three linguistic factors corresponding to different levels of analysis were tested:

– the final sounds of words;

– their last letters, defined as the spelling reproducing the last vowel and, if applicable, its corresponding coda, for example, *-one* in the word *trombone*;

– the suffix, in the cases where there was one.

This study made it possible to show that the grammatical gender of a noun can be predicted in a large number of cases, contrary to the claims of many French theoretical studies in grammar. Furthermore, the spelling's ending seems to be the best of the three predictors. In the French language, certain sequences of letters are strongly associated with a certain grammatical genre, for example, *-ette* is associated with the feminine form, regardless of whether this is a suffix, as in the word *statuette* (small statue), or not, as in the word *devinette* (riddle). For this study, the use of a word list offers major advantages compared to a corpus: the grammatical category of words is already known, which simplifies noun retrieval: every word is already associated with its grammatical gender. These two pieces of information are missing from raw corpora. What is more, since all nouns in French have a grammatical gender (or in rare cases, even two), a dictionary offers an extensive list of examples which help us look for regularities.

The use of natural data from corpora rather than simple word lists is nonetheless essential to answer other research questions in morphology. First, in order to assess the productivity of a morphological rule, or even to assess word formation, that is, whether certain words made up from derivation or morphological composition exist, a corpus can offer much more information than the one found in a dictionary. Indeed, these corpora, and in particular, the large corpora available via the Internet, make it possible to find occurrences even for very rare words, as well as to provide very recent examples of language uses. This is all the more significant at a moment in the history of language when these uses have not yet been listed in dictionaries. The use of corpora also offers the possibility of finding out the context in which a certain word was produced, with the aim of checking, for example, whether the meaning of the derived word in such a context was the one expected. Corpus-based research has also made it possible to show that certain derivations deemed impossible from the point of view of theoretical morphology did nonetheless exist. For example, in theoretical analyses of French morphology, it was deemed impossible to attach the prefix *anti-* to a morphological pattern such as *verb* + *-able* (Fradin 1997). Despite this, occurrences such as *anti-inflammable* have been identified in corpora retrieved from the Internet (Hathout *et al.* 2008).

Second, in their study focusing on the derivation *-able*, Hathout *et al.* (2003) showed that this form of derivation has been applied in significantly more cases – and with much more diverse meanings – than those expected on the basis of word lists. In their study, they looked for cases on the Internet trying to identify words derived in *-able*, using Webaffix, a tool that allows us to retrieve morphemes. This research enabled them to study a list of about 5,000 occurrences, whereas previous studies, based on word lists, had been limited to about 1,400 occurrences. Previous studies had concluded that the derivation *-able* was essentially a means of forming verbal adjectives with a passive meaning, insofar as they serve to modify nouns which are typically in the position of the direct object of the basic verb modified by the suffix. For example, *buvable* (drinkable) can be used to turn the noun *boisson* (drink) into *une boisson buvable* (a drinkable drink). The noun *boisson* (drink) is often the direct complement of *boire* (to drink), as in *boire une boisson* (to drink a drink). The manual analysis of data from the Internet has shown that the uses of *-able* are actually much more diverse and also apply to noun bases, for example, *un terrain piscinable* (literally a ground in which a swimming pool can be built) or *une statue muséable* (literally a statue worthy of being put in a museum).

Finally, the use of corpora makes it possible to identify whether certain morphological derivations are more or less productive in certain language registers, or if they are produced more by certain types of speakers (e.g. within a determined age group) or in geographical areas. However, word lists do not make it possible to answer such questions, since they represent a normative use, corresponding to a standard variety. Corpora also help to trace how the productivity of a morphological rule evolves over several decades by means of monitor corpora, or even over several centuries, by comparing corpora from different periods (Baayen 2008).

The study by Koehl and Lignon (2014) regarding adjectival nominalizations such as *-ité* and *-itude* perfectly illustrates this point. Since the 16th Century, French dictionaries have recorded very few new *-itude* suffixes. This suffix no longer deemed productive. By comparing data from a word list drawn from a computerized version of *Trésor de la Langue Française* (a major online French dictionary), data from the newspaper corpus *Le Monde* and an Internet search, the authors were able to prove that the suffix *-itude* had experienced a strong recent upsurge, but that the latter was only evident on data drawn from the Internet. Thanks to the numerous occurrences drawn from the two corpora, they were also able to compare the meanings of the suffixes *-ité* and *-itude*. Their results showed that these suffixes, despite being close, each had their own specificities. On the one hand, *-itude* tends to denote an attitude, whereas *-ité* tends to denote a concrete object or a relation. In summary, including data drawn from the

Internet has revealed a recent development, associated with a more informal language register than newspaper articles or literary works.

2.3. Syntax

The use of corpora in the field of syntax has been controversial for a long time, since this field has relied on an essentially rationalist methodology for several decades, as a result of Chomsky's works (see Chapter 1). However, the use of corpora in the field of syntax has grown considerably. In particular, the use of corpora makes it possible to compare the productions of various speakers, of different language varieties as well as different registers, providing a much more nuanced and realistic vision of the structures underlying language uses, rather than the intuitions of a single speaker. In addition, the syntactic analyses of corpora have made it possible to obtain a fine measurement of the frequency of so-called grammatical sentences, compared to those considered as ungrammatical, and thus making it possible to overcome such binary opposition. This frequency analysis can also be completed by an analysis of other factors (lexical, grammatical, discursive, etc.), making the uses of certain types of constructions more or less likely to occur in various discourse genres. The large amount of data provided by corpora, combined with the use of automatic analysis tools, have also made it possible to uncover trends which could not have been observed with the naked eye on the basis of just a few occurrences.

The study by Verwimp and Lahousse (2017) has illustrated how the study of corpora makes it possible to identify new semantic functions associated with certain syntactic structures. Their analysis focuses on cleft constructions such as those introduced by "*il y a*" (there is/are) as in (2).

(2) *Dans la rue, il y a des femmes qui discutent.*

(In the street, women are talking.)

In the literature, this type of structure is associated with the presentation of new events in discourse. By performing a corpus study on spoken French using the *Corpus de français parlé parisien des années 2000*, the authors identified all the occurrences of "*il y a*" or "*y a*" (both forms meaning "there is") and then manually chose only the cases (98 in total) in which *(il) y a* was followed by a definite noun phrase and a pseudo-relative. Among these sentences, only 16 occurrences corresponded to cases where a new event was being introduced in the discourse, as the literature claims. Analyzing examples helped the authors identify other types of functions for this structure, notably the introduction of a new entity in discourse, as in (3).

(3) *Il y a Sophie qui veut te parler.*

(Sophie wants to talk to you.)

These different functions were more easily identified thanks to the large availability of contextual language in corpus data. This study illustrates how a corpus study can combine quantitative elements (the prevalence of different functions for a structure) with qualitative ones (the identification of semantic functions).

Verwimp and Lahousse's study could be carried out on a transcribed spoken corpus in the absence of any form of syntactic notation, since the structure the authors were looking for matched a clearly identified lexical pattern. Nevertheless, the same does not apply to the study of other structures that are not associated with such a pattern. For this type of case, the use of a syntactically annotated corpus becomes compulsory. This is illustrated by the second study that we will discuss in this section.

Fabre and Bourigault (2008) used a corpus annotated with a part-of-speech tagger and a syntactic parser, that is, a corpus which had been classified into grammatical categories and grouped into phrases and dependencies in order to study the alternation between prepositional phrases functioning as verbal arguments or attached at the sentence level. For this endeavor, the authors used a sub-section of the *Frantext* corpus containing 520 novels from the 20th Century, which they had syntactically analyzed via the Syntex software. Based on this corpus, they calculated the association force between verbs and prepositions, namely taking into account factors such as the diversity of contexts in which these associations are found, which makes it possible to calculate the productivity of such a pair. Indeed, the more often a given verb and prepositional phrase association is found in different contexts, the more it means that a verb is regularly associated with a certain proposition. Afterwards, they calculated the degree of autonomy of the "preposition + noun phrase" groups, based on the number of different verbs with which they were used. These criteria helped the authors calculate an autonomy coefficient for each prepositional phrase. The results of this study have shown that certain prepositional phrases are used in recurrent combinations with certain verbs and therefore seem to be linked to the semantics of the verb, whereas other phrases seem to be more autonomous from the verb. The main interest of this study is to have proved the existence of a continuum in the attachment force of prepositional phrases to verbs, paving the way to a much finer analysis of this phenomenon than the one offered by the binary distinction between verbal and phrasal arguments, which can be problematic in many cases.

Another large-scale annotated corpus study was carried out to explore the question of the placement of the attributive adjective, either before or after the noun it modifies. In French, this is considered a complex question, because the factors leading to one placement or the other are linked to phonology, morphology, syntax, semantics and discourse. Among all these factors, the identification of those playing the most important role cannot be carried out without a quantitative analysis of large-scale data. Thuilier *et al.* (2010) attempted to identify these factors using syntactically annotated corpora. Based on the literature, the authors identified 13 variables that could influence the type of placement, such as adjective length (in syllables), whether or not it is morphologically derived, and its frequency in the corpus. Using statistical analysis, the authors were able to establish which of these factors best predicts the adjective's position. Their results indicated that the factors related to language use, such as adjective frequency and the frequency of the contexts in which it appears (its collocations), are the main factors for predicting the placement of an adjective, rather than variables related to the linguistic system.

In conclusion, the analysis of syntactic phenomena using large annotated corpora helps us go beyond traditional qualitative and rationalist analyses. However, these analyses are only possible on corpora that have been parsed syntactically, and these are still rare due to the complexity of applying automatic parsers. Spoken data are particularly difficult to automatically annotate syntactically due to the presence of numerous repetitions, hesitations, etc., which ruins parsing. The results from studies on syntactically parsed corpora cannot therefore be generalized to all language registers for the moment. However, it is likely that these studies will increase in the future insomuch as the quality of syntactic parsers will keep on improving.

2.4. Lexicon

Unlike syntax, lexicon is ideally suited for corpus analysis, and it is also the field that has been approached from that perspective for the longest time. In fact, most of the time, the study of lexicon can be done directly on a raw corpus, or in some cases, on a corpus that has been annotated with part-of-speech tagging.

Corpora make it possible to identify all the words used in a language as thoroughly as possible. It is evidently impossible to list all the words existing in any language, but using large corpora, it has become possible to get a much more realistic idea of the number of words in circulation compared to the lists that could be drawn from dictionary databases, which list only part of it. For example, based on the *Google Books* corpus, which contains more than 500 billion words (including 361 billion in English), drawn from literary sources from the 16th Century to the present day, Michel *et al.* (2011) were able to estimate that the English lexicon in the 2000s was made up of about 1,022,000 words. This figure implies that more than 50% of the English lexicon is not listed in any dictionary, because dictionaries are only limited to a few hundred thousand words, even the most exhaustive ones. By comparing the evolution of lexical diversity over time, the authors were also able to show that the English lexicon has greatly developed since the 1950s, with an average of 8,500 new words (including proper and compound nouns) being added every year. In addition to making it possible to identify a language's lexicon, lexical studies using corpora also help determine the most frequent words belonging to the basic lexicon of a language. These studies thus make it possible to estimate the difficulty of a word and to adjust the teaching material intended for children and learners accordingly (see Chapter 3).

Corpora also have another great advantage for lexicon study: they make it possible to identify the sequences of words that frequently appear together, which are called collocations, for example *prendre un douche* (to take a shower) or *forte pluie* (pouring rain). We can thus identify fixed sequences like idioms, as well as determine their degree of fixedness. For example, these studies help us to establish whether it is possible for a certain idiom to be used in the passive form or not, with different verbal tenses, or whether the idiom's topic is fixed, as in *j'en mettrais ma main au feu* (I would stake my life on it), or free, as in *mettre le feu aux poudres* (to stir up a hornets' nest).

From the point of view of lexical meaning, corpora are valuable resources for determining the vast array of meanings that a word may take on in different contexts. By means of corpus studies, it is also possible to define lexical fields and to study meaning relations between words, such as synonymy and anonymity. For example, in order to determine whether two words are synonymous or not, the analysis of corpus data makes it possible to determine whether these two words can appear in the same linguistic context or not. Finally, the analysis of corpora from different textual and spoken genres makes it possible to determine the situational contexts in which the words are used, in a much more nuanced way than register indications, such as *soutenu* (formal) or *populaire* (informal), found in dictionaries (see Chapter 3, section 3.5).

The first example of a study that we will introduce in this section is an antonym analysis and, more specifically, antonyms which are used together within the same utterance, as in (4), where the use of antonyms in bold serves to reinforce the contrast with another pair of opposites in the utterance, in this case, between students and workers.

> (4) The initiative was very **popular** with the students and very **unpopular** among workers.

Steffens (2018) studied the use of these antonyms on the basis of a journalistic style corpus drawn from the archives of *Le Monde* newspaper. Using the list of antonyms appearing in *Le Grand Robert* dictionary and their derivatives, the author was able to compile a list of 35,000 pairs of antonyms, which she then retrieved from the corpus using tools that make it possible to perform searches of several words within the same sentence. This research enabled her to show that functions similar to those identified in English by Jones (2002) were also found in French and that a detailed qualitative analysis of these examples made it possible to reconsider this classification based on semantic and pragmatic criteria, rather than on their linguistic form. This study illustrates how the use of large corpora helps to identify the numerous occurrences of a phenomenon, to analyze such occurrences within a rich context and to reveal the communicative strategies underlying them. Although the author carried out a qualitative analysis on the basis of her corpus, these data would also have been suitable for a quantitative study regarding the prevalence of each of the functions identified.

The second lexicon study that we will introduce deals with the question of neologisms and, more specifically, how they get into the language or not. In France, the question of neologisms is a sensitive point, especially when the new words have been borrowed from English. The rejection of Anglicisms gave rise to several decrees, and then to the law concerning the use of the French language (known as the Toubon law), which prohibits the use of foreign terms in public administration texts. One of the consequences of these decrees was the creation of a Commission for the enrichment of the French language (formerly known as the Terminology and Neology Commission), which was responsible for suggesting French terms that could help avoid Anglicisms. In order to study whether these suggestions are actually used by French speakers, Berthelet (2015) drew word pairs from these lists containing, on the one hand, the word suggested by the Commission and, on the other hand, the Anglicism, which it was supposed to replace (e.g. *parrainage* instead of *sponsorship*, or *numériseur* instead of *scanner*), limiting the analysis to the fields of sport and communication. For each French word, the author also wrote down the year in which it had been proposed by the Commission. He then looked for occurrences of these doublets in the French-speaking section of the above-mentioned *Google Books* corpus. The results clearly revealed the ineffectiveness of such measures. Either the French word was much less used than its borrowed equivalent and this trend did not change after the introduction of the suggestion made by the Commission, or the French word was indeed the most frequent term, long before it had been legitimized by the Commission. For example, this is the case of words such as *logiciel* (software), which French speakers have adopted long since.

Finally, we will mention another study concerning the French lexicon and its connections with English, which was conducted in the context of French-speaking Canada. In Canada, the question of Anglicisms is also sensitive, insofar as French is a minority language and struggles to keep its importance alongside English. Planchon (2018) wanted to determine whether Anglicisms were more used in the written press of certain French-speaking regions compared to others. To do this, she consulted the online archives of three local newspapers from different regions (Montreal, Ottawa and Quebec). Based on the computerized version of the *Multidictionnaire de la langue française*, she retrieved a list of criticized borrowings (495 in total), which she then looked up for in each of the newspapers. In order to avoid data bias, she was careful to exclude articles taken from other newspapers, as well as translations. The results indicate that the newspaper from the region where French is the minority language (Ottawa) is also the one with the lowest number of Anglicisms. This result tends to reinforce the hypothesis that the regions where French is the minority language are the same where speakers tend to defend it the most. However, this conclusion was drawn only from descriptive data and may have benefited from being reinforced by the use of statistical tests, which we will discuss in Chapter 8.

2.5. Discourse analysis

The field of discourse analysis is an empirical field *par excellence*, insofar as the objective is to study the structure and content of different types of interactions. However, for a long time, these studies were carried out in an almost exclusively qualitative manner, with the aim of analyzing certain particular situations in detail. These studies are very useful for understanding the nature of interactions. For example, Traverso (2019) studied the structure of requests from people asking for help in an office with access to social rights. This study provided a very detailed analysis of the linguistic, paralinguistic and gestural resources used in this particular situation. All these criteria cannot be contemplated simultaneously by a quantitative analysis. Discourse analysis studies also triggered the development of the first multimodal corpora, which make it possible to study interactions in very rich contexts. In addition to these qualitative studies, it is also possible to analyze certain aspects of discourse from a quantitative perspective. We will focus on this type of analysis in the following section.

The discourse elements that are best suited for a corpus quantitative analysis are those easily identified on the basis of raw data. For example, many studies have focused on how discourse markers such as *bon, ben* and *alors* (well, so, I mean, etc.) are used in different kinds of speech and with what functions (e.g. Crible 2018). The example of *bon* (well) illustrates that many markers are polysemic (in this case, as a marker or as an adjective) and require a manual analysis of data in order to identify relevant occurrences. Other quantitative studies have analyzed the way in which speaking turns change (e.g. Beňuš *et al.* 2011), both from the prosodic point of view and from the structuring of discourse (hesitations, reformulations, etc.).

The above-mentioned studies mainly apply to spoken and interactive language. However, other studies have also focused on written discursive genres. In particular, these studies aim to study the way in which discourses are constructed from the point of view of their cohesion, by analyzing discourse connectives such as *parce que, donc* and *quand* (because, therefore and when), for example in Degand and Fagard (2011) and Zufferey and Degand (2017), as well as by studying referential expressions and pronominal anaphora, more specifically. In this case, the extraction of corpus data should also be combined with a manual selection of occurrences, since the relevant cases cannot be retrieved directly from the raw data. The manual annotation of a limited number of occurrences still provides a solid basis for statistical analysis, as we will see below.

Quantitative discourse studies have also sought to study the differences between discursive genres (Biber and Conrad 2019), focusing on the way they are structured, the use they make of connectives and of discourse markers or the linguistic specificities of each genre, depending on whether they are monological or dialogical, planned or spontaneous, spoken or written, formal or informal, etc. For example, Crible (2018) compared the way in which discourse markers are used in a corpus comprising seven different discourse genres and showed that the number of markers produced – as well as their functions – vary considerably depending on the characteristics of the different genres.

The study by Simon and Degand (2007), which we will discuss for illustrative purposes, deals with the semantic differences between the two French causal connectives *car* and *parce que* (since and because), as well as their different spoken and written uses. Based on the literature, the authors hypothesized that the semantic difference between *car* and *parce que* stems from the degree of subjectivity that is expressed in the causal relationship. While the connective *parce que* is typically used for expressing objective causal relations between states or events in the world as in (5), the connective *car* is typically used for expressing more subjective relations, which connect a statement or a conclusion drawn by the speaker and its justification, as in (6).

(5) *Emma est arrivée en retard parce qu'elle a raté son train.*

(Emma arrived late because she missed her train).

(6) *Emma est très désorganisée, car elle perd tout le temps ses affaires.*

(Emma is very disorganized because she loses her stuff all the time).

Simon and Degand further hypothesized that the difference between *car* and *parce que* is not stable between the spoken and written modes, due to the fact that the connective *car* no longer seems to be widely used in spontaneous spoken discourse. In order to verify these hypotheses, they used a written corpus (*Le Soir* newspaper) and a spoken corpus (Valibel database), both representing the variety of French spoken in Belgium. First, they retrieved all the occurrences of *car* and *parce que*. They found that in written language, the two connectives have a very similar frequency (approximately one occurrence every 300 words for *parce que* and every 250 words for *car*), whereas their frequency is quite different in spoken language. On the one hand, *parce que* is 10 times more frequent in spoken language than in written language, whereas it is the opposite for *car*, which is 10 times less frequent than in written language. In spoken language, *parce que* is more than 185 times more frequent than *car*.

The authors then manually classified 50 occurrences of each connective found in the spoken and written modes as either objective or subjective, that is, a total of 200 occurrences, randomly chosen from the corpus. Then, for each connective, they were able to compare the differences in distribution between the types of relations both in the written and spoken modes. The results indicated that in written language, *car* is a more subjective connective than *parce que*. However, in spoken language, *parce que* is used for communicating all types of causal relations, replacing *car*. This study shows the semantic criteria for making a distinction between two connectives that have a similar meaning and reveals important differences in their use between the spoken and written modes.

The second study we will discuss in detail in this section deals with a particular textual genre, namely the SMS language and its influence on the young generation's command of other written genres. Cougnon *et al.* (2017) tested the ability of 80 young people aged between 14 and 15 years to make a distinction between discourse genres when they write. More specifically, they sought to determine whether there is a link between the way young people write on social media and their command of spelling. To do this, they created three small corpora comprising Facebook conversations, written essays and two dictations. Some students took part in only one or two of these activities. In total, data comprising the three discourse genres were collected for only 10 students.

The authors analyzed the types of spelling mistakes produced by the students. For the dictation exercise, they found that the grammatical errors were the highest (agreement, etc.). In Facebook conversations, spelling alterations to typical written words in social media represented less than one in three words, which contradicts the idea that social media language is entirely different from other language registers. Next, the authors showed that there is no link between the propensity of students to use alterations on social networks and their spelling skills in the two dictations. The authors also investigated whether the formal alterations found in the language of social networks had repercussions in other discursive genres. Unfortunately, only a list of 30 words had occurrences in the different genres, which is insufficient for carrying out a truly quantitative analysis. However, they could see that the altered forms in social network conversations (*bcp* instead of *beaucoup*, *c* instead of *c'est*, etc.) were not found in the spelling of these words in the other register nor were they misspelled in the dictation exercises. This study thus offers arguments against the misconception that the writing style in social networks degrades young people's spelling skills and makes them unable to differentiate textual genres.

2.6. Pragmatics

Pragmatics studies language use in context. This definition brings together a wide range of heterogeneous phenomena such as speech acts, implicatures, politeness phenomena and conversation analysis. Pragmatics has many points of contact with both discourse analysis and sociolinguistics. As in the case of these two disciplines, corpora represent valuable tools in pragmatics, because they make it possible to study the use of language in real communication situations.

Certain pragmatic phenomena such as turn-taking in conversations or the use of discourse markers discussed in the previous section can be studied by looking for certain linguistic forms, for example discourse markers such as *bon, ben, voilà* (well, so, actually) or certain specific parts of corpora, such as the first utterances in conversations, depending on whose turn it is to speak. However, for other pragmatic phenomena such as speech acts and implicatures, as well as for expressing politeness, there is no systematic relationship between linguistic forms and pragmatic functions.

For these phenomena, it is necessary to use data annotation and this annotation must be done manually on the basis of an observation of the entire corpus. This process may become time-consuming, and annotations are not always easy to carry out, insofar as the speech acts that speakers communicate by means of their utterances in many cases are not communicated transparently. For example, it is possible to ask someone to open a window with a formulation such as *Could you open the window?* which explicitly mentions the subject of the request. However, this can also be done by means of much more indirect formulations such as *It's hot in here!* or *It's hard to breathe*. The role of this type of formulation largely depends on the context and cannot be automatically inferred from the linguistic form employed. As soon as a speech act is associated with a certain type of utterance (e.g. interrogative sentences) or frequently associated with certain words (such as *sorry* or *oops* for excuses), a corpus search becomes greatly simplified. This research should nonetheless be subject to manual verification (see Jucker (2009) for a more detailed discussion of this question). It is for this reason that the use of corpus data has been limited to certain areas of pragmatics, but this situation is evolving as best practices for the annotation of pragmatic phenomena gradually develop (Rühlemann 2018).

We have already illustrated the use of corpora for the study of discourse markers and connectives in the previous section. In this section, we will introduce a study illustrating the usefulness of corpus data for the study of scalar implicatures. These implicatures are communicated through the use of a weak scalar term which contextually excludes the affirmation of the stronger term. For example, the quantifier *certains* (some) as in (7) pragmatically excludes the interpretation, although logically possible, according to which all of Laura's friends are nice.

(7) *Certains amis de Laura sont sympathiques.*

(Some of Laura's friends are nice).

Thus, the use of quantifiers generates a pragmatically enriched interpretation in the form of an implicature: *certains mais pas tous* (some friends, but not all of them, are nice). Some pragmatists (e.g. Levinson 2000) believe that quantifiers may give rise to generalized implicatures; in other words, these are generated regardless of the context of the utterance. This phenomenon has been the subject of a very abundant literature in theoretical and experimental pragmatics (see Zufferey *et al.* (2019) for a review). Most of the experimental results question the existence of generalized implicatures generated by default, in the absence of contextual information. However, these results are all based on utterances especially designed for the needs of experiments, rather than real uses drawn from corpora, which calls their naturalness into question.

Degen (2015) came up with the idea of using corpus data to obtain a real set of uses for the quantifier *some* in English (the equivalent of *certains*), in order to experimentally test whether native speakers derive the implicature *some but not all* in every case as predicted by the theory of generalized implicatures. In the corpus of telephone conversations (switchboard), the author retrieved all the occurrences of *some* followed by a noun phrase, and after excluding certain problematic cases, kept 1,363 occurrences for analysis. She asked a large group of participants to read all the occurrences of *some* and to tell her for each case whether a reformulation containing the meaning derived by implicature, for example, *Some but not all of Laura's friends are nice*, properly matched the speaker's meaning. Thanks to this method, she was able to show that not all uses of the quantifier gave rise to an implicature, contrary to the hypothesis of generalized implicatures. She was also able to identify the linguistic and discursive factors that are linked to the derivation of an implicature on the basis of the analysis of the linguistic contexts regarding these uses of the quantifier. The use of corpus data also helped the author analyze numerous real examples framed against a vast context, which is not the case with the experimental material, often compiled *ad hoc* for the needs of a study.

2.7. Sociolinguistics

In order to study social variations in language use, sociolinguistics necessarily resorts to external data. The use of corpora has therefore long been a fundamental tool in sociolinguistics. In particular, it makes it possible to compare how different social groups such as women, men and people in the cities, the suburbs or in the countryside use language in natural situations. Another major concern of sociolinguistics is to document regional variations in the use of languages, and for this, corpora are also an essential resource. Moreover, numerous corpora have been developed with the aim of documenting the use of certain regional varieties such as French-speaking Switzerland, Belgium and Canada (see Chapter 4). Sociolinguists also believe that variation is a sign of an ongoing change in language, and in many cases, corpora represent an effective tool for uncovering this kind of variation, particularly in the different registers of the spoken language.

The first study we will introduce deals with the issue of linguistic changes. Gardner-Chloros and Secova (2018) studied the formulation of indirect questions in Parisian French and, more specifically, the existence of indirect questions in which the interrogative word is positioned *in situ* in embedded structures as in (8), rather than the standard variant (9).

(8) *Je ne sais pas il fait quoi.*

(literally: I don't know he does what).

(9) *Je ne sais pas ce qu'il fait.*

(I don't know what he is doing).

In order to study the distribution and prevalence of this type of structure in the French spoken in Parisian suburbs, all the occurrences of indirect questions were collected in a small oral corpus of approximately 350,000 words. The identity of the person who produced the occurrence was then classified into different categories:

– age group;

– ethnicity (French parents, mixed origin, two immigrant parents from the same culture);

– the diversity of friends' networks (in terms of percentage of friends in the same ethnic group);

– degree of bilingualism (French and another language).

This coding enabled the authors to count how many occurrences were produced following the different criteria identified and to prove that young people from bilingual backgrounds use the post-verbal structure (8) significantly more when compared to young people from monolingual French-speaking families. The ethnic group also plays an important role, since young people with two immigrant parents also use this structure significantly more than young people from French families. Gender also plays a salient role, since boys tend to use such structures more than girls. Finally, having a culturally diverse network of friends is also correlated with the use of these structures: young people with a group of friends mixed at 80% tend to use them more than young people with an unmixed group, or a group mixed at 20%. The authors also tested which of all the factors mentioned above best predicted the type of indirect question used (pre-verbal as in (8) or post-verbal as in (9)). The results indicate that the best predictor is ethnic origin, followed by gender and finally by the degree of diversity in their friends' network.

The authors finally compared the results of their corpus with the use of these same structures in the *Corpus de français parlé parisien des années 2000*. They observed that this structure was much less used in this second corpus, with only two occurrences by male speakers of Moroccan origin. Thus, this study showed that the use of post-verbal indirect questions represents a case of language change initiated by the less privileged social strata of the population, rather than a prestige change (as is the case with other sociolinguistic changes). This change seems motivated by the desire to give more weight to the interrogative word by placing it at the end of the utterance, while keeping the same word order *in situ* in direct questions and in embedded questions.

Bernicot *et al.* (2012) also studied the role of social factors such as gender and age, but this time on how to write text messages. These authors studied various criteria such as message length, internal structure (with or without opening and closing) and their main communicative function (communicative or relational). This study was based on the corpus of SMS (*Sms4science*), for which they chose 91 monolingual French-speaking participants. These participants were then divided into groups depending on three variables:

– age (15–16 or 17–18 years);

– gender;

– the level of experience in writing text messages, the latter variable being coded in a binary way (experienced or inexperienced), from the number of messages sent each day.

Each of the messages in the corpus was coded according to the three linguistic variables to be studied, namely the message's length, the presence of opening and closing elements and its main function. The results indicated that text messages differ from other text genres in that 73% of them do not contain an opening and/or closing expression. As regards the effect of the authors' gender, the messages written by girls tend to be longer than those of boys, but only among the age group of 15–16 years. The level of experience also has an influence on the type of message produced. Inexperienced writers most frequently produce text messages with opening and closing expressions. Finally, SMS mainly has a relational function more than an informative one (56% of messages), and this trend tends to be more pronounced in the younger group and among girls. This study illustrates the way in which different social factors influence textual production within the context of a particular discursive genre.

2.8. Diachronic linguistics

The quantitative study of the linguistic changes that languages experience over the centuries, is based on the study of corpus data. Such data makes it possible to document the different stages of these changes, as well as the different periods in the history of a language in which these changes took place. The main limitation to the use of corpus data for studying linguistic changes from a diachronic perspective lies in the fact that linguistic changes generally first take place in the spoken language. However, the first spoken corpora date from the second half of the 20th Century. For previous periods, data is limited to written records and, more specifically, to certain registers such as literature and legislation. Due to the lack of spoken records, linguists sometimes resort to the speaking attributes of fictional characters in dialogue as a source. While these dialogue excerpts reveal a less formal language than in other written genres, they do not reproduce certain essential characteristics of the spontaneous spoken language, such as hesitations and errors. Certainly, these data offer no indications concerning phonology or prosody, which are nonetheless important elements in the processes of linguistic change.

The first study that we will introduce in this section deals with the emergence of the counterfactual conditional in structures as shown in (10).

(10) *Si j'avais su, je ne serais pas venu.*

(If I had known, I would not have come).

In order to trace the emergence of this function of the conditional in French, Patar *et al.* (2015) compiled a diachronic corpus of works in French from the 11th to the 20th Century, totaling approximately 9.8 million words. The texts included in this corpus were drawn from diverse textual genres, such as literary texts, non-fictional narrative genres (memoirs, journalistic texts, etc.), argumentative texts, correspondence and, finally, a "miscellaneous" category made up of texts which do not fall into these categories (e.g. legal texts). Even if the proportions of each sub-corpora were not equivalent, the authors were careful to include a similar distribution for each of the periods analyzed. The authors then retrieved all the occurrences of the auxiliaries *être* (to be) and *avoir* (to have) in the present conditional and identified 90 possible spellings, which they then looked up in their corpora in order to identify the occurrences in the past conditional. For the last four centuries of the corpus including an enormous number of occurrences, the authors randomly selected 100 occurrences per century, in order to obtain samples of comparable size for all the periods. Based on the literature, the authors were then able to identify five possible functions for the past conditional, including the counterfactual conditional, and annotated every occurrence in their corpus depending on one of these five functions.

The results indicated that the first evidence of the past conditional goes back to the 12th Century but that the frequency of this structure was fairly low until the 16th Century, with around 10 occurrences every 100,000 words. The frequency then sharply increased from the 17th Century and reached 70 occurrences every 100,000 words from the 18th Century onwards. The moment when the uses of the past conditional increased in the 17th Century matched a strong progression in its counterfactual uses, which became the most broadly employed function of this construction. Linguistic theories on the evolution of perfect forms predict that evolution can develop from a previous interpretation (e.g. when the process described by the verb, namely the action, the state or the result described, precedes another process in the sentence) to a past interpretation (when the process described by the verb represents a bygone situation at the time of the utterance). The authors therefore coded every occurrence according to the type of process described: previous or past. Their results showed that the past conditional had an essentially previous interpretation until the 17th Century, when the past interpretation began to increase. This change matched a strong growth in the counterfactual uses of the past conditional. The study thus illustrates how corpora make it possible to document the emergence and the later evolution of a certain language trait. Thanks to the annotation performed on data themselves, it also shows that certain frequency changes can be associated with the emergence of new functions.

The second study that we will discuss concerns the pair of causal connectives *car* and *parce que*, which we have already mentioned above (section 2.5). Degand and Fagard (2012) studied the diachronic evolution in the use of these two connectives. To do this, they worked with the *Frantext* database, choosing narrative texts from the pre-classical period (from 1550) when *parce que* became stable as a subordinating conjunction (but not yet completely grammaticalized in its current form), until its uses in contemporary French. They then looked for the occurrences of these connectives in the corpus and chose 100 occurrences of each connective for each period of the French language evolution. They annotated all of these occurrences according to the type of relation communicated: objective or subjective (in this study, the authors divided these categories into several subcategories, which we will not discuss here). The results indicated that in general the connective *car* is used significantly less than *parce que* to describe objective causal relations. In addition, over time, the number of subjective causal relations significantly increased in the corpus between Middle French and modern French. Ever since the Middle French period, *parce que* has been used in subjective causal relationships and these uses have increased over time. Conversely, since the beginning and still nowadays, *car* is a connective used primarily for expressing subjective causal relations. It is possible that this evolution of *parce que*, which went hand in hand with a strong increase in the number of occurrences (from 7 occurrences every 10,000 words in Middle French to 37 in contemporary spoken French), is the reason why *car* has now almost completely disappeared from the spoken language, as we have already seen. This study also illustrates the links between the quantitative evolution in the use of a connective and the evolution of its functions.

2.9. Conclusion

In this chapter, we have shown that the use of corpora can prove to be a valuable tool in all areas of theoretical linguistics. One of the main advantages of corpora is that they contain natural data providing a glimpse of different forms of language use, which can thus be studied while taking into account a rich linguistic context. We have seen that certain studies can be carried out on the basis of raw data, particularly in the field of lexicon, whereas in other fields, such as syntax or discourse analysis, usually data has to be annotated, something which can be done manually or partially automated. We will discuss the topic of manual annotations in Chapter 7. The studies introduced in this chapter have also shown that it is possible to quantitatively analyze different types of observations and data. We will introduce the most accessible methods in detail in Chapter 8. Finally, all the corpus studies illustrated in this chapter were carried out on corpora of very diverse shape and size. In Chapter 5, we will discuss in more depth the data that are available in French to the public for carrying out corpus studies.

2.10. Revision questions and answer key

2.10.1. *Questions*

1) What data are necessary to be able to conduct a corpus study on the way the pronunciation of vowels like the [a] in *patte* and the [ɑ] in *pâte* has evolved in French-speaking Switzerland and in France?

2) Why is the use of word lists not always enough for studying morphological phenomena in a language?

3) Why is syntax a field of study in linguistics where the use of corpora is still limited?

4) Which are the lexicon studies that cannot be fully performed on raw data?

5) What are the areas of discourse analysis properly suited to carry out quantitative studies?

6) What are the constraints posed by the study of pragmatic phenomena such as speech acts and implicatures on a corpus study?

7) Which are the sociological factors suitable for carrying out the study of linguistic phenomena on corpora? What types of data do we need to perform these analyses?

8) What are the limitations to the use of corpora for studying the evolution of languages diachronically?

2.10.2. *Answer key*

1) In order to study this question, it is necessary to use two spoken corpora respectively representing a variety of French spoken in France, for example in Paris, and that of French-speaking Switzerland. These corpora must also contain diachronic information, for example regarding the evolution of pronunciation from the second half of the 20th Century to the present day. A selection of occurrences of the vowels to be studied, representing the different periods, should then be looked up in the corpus. For each occurrence, the pronunciation should be noted, either by a phonetic analysis of the production of the vowel or by a manual annotation carried out by two native speakers. Then, the various pronunciations should be quantitatively compared between the two corpora and the periods involved.

2) First, when using word lists, morphologists do not have access to the linguistic context in which these words were produced. For a number of research questions, this type of information is crucial, for example, when it comes to determining the meaning of a derived word. Second, by using a word list from a dictionary rather than a corpus, morphologists only have access to a limited subset of the speakers' linguistic productions. In fact, dictionaries provide a list of well-established uses in language, which correspond to the language's standard variety. However, new uses are constantly emerging and these are not listed in dictionaries. Likewise, corpora also represent less formal varieties of language than the written standard found in dictionaries, which makes it possible to diversify the language uses identified in this way. Finally, large corpora make it possible to have access to much more data than word lists, which makes it possible to identify rare phenomena which do not appear in dictionaries.

3) Syntax has been the area of language most influenced by Chomsky's work and has followed for decades the rationalist methodology that this author advocated for linguistics. Indeed, formulating grammaticality judgments is one of the areas in linguistic studies for which native speakers can trust their intuitions, albeit partially. At present, this objection in principle is largely over, but the main remaining barrier is of a methodological nature, since the study of many syntactic phenomena requires the use of syntactically annotated corpora. However, syntactic annotation tools are still imperfect and sometimes too complex to implement or use. As a result, few corpora have been the subject of such annotations and their complexity discourages certain linguists from using them.

4) Unlike syntax, in many cases, the lexicon can be studied on raw data, that is, non-annotated data. However, due to the existence of many polysemic and homonymic words in French, many lexical searches must be done on the basis of morphosyntaxically annotated data in order to avoid identifying parasitic occurrences. For instance, when looking up the word "*orange*" as a color, it is necessary to separate its adjectival uses from its nominal uses. In addition, research on the metaphorical uses of a word must be subject to manual annotation on the basis of their context, since metaphors have no lexical markers that make it possible to differentiate them from the literal uses of the same words.

5) In general, all the elements that can be identified on the basis of surface lexical features are well-suited for quantitative analysis, since it is possible to identify numerous occurrences by means of a computerized search. We should nonetheless observe that this first identification often requires a later manual sorting of the data. In the case of discourse, the phenomena which are well-suited for quantitative analysis are the lexical cohesion markers. These are the use of anaphoric pronouns (*he*, *she*, etc.), discourse connectives indicating coherence relations between discursive units (*because*, *when*, *if*, etc.) and the use of discourse markers (*well*, *I mean*, *you see*, etc.), which index the interpersonal relationships between the speakers and offer clues to the discursive planning of the speakers. Conversely, phenomena such as the structuring of information within discourse and the analysis of global coherence are much more difficult to annotate on a large scale since they require an entirely manual processing of the corpus.

6) The main constraint posed by many pragmatic phenomena such as the study of speech acts and implicatures is related to the fact that the occurrences of these phenomena cannot be identified on the basis of lexical markers which can be automatically searched for in the corpus. For example, an indirect speech act such as a request can take many different superficial forms as a question (*could you shut up?*) or an assertion (*I'd like you to stop talking*). The same goes for implicatures, which are not linked to the use of specific words, with the exception of generalized scalar implicatures, which are typically associated with the use of quantifiers such as *some* and *a few*, logical connectives like *or* and some telic verbs such as *to start*. In order to identify the occurrences of pragmatic phenomena which are not related to words, it is therefore necessary to scan the entire corpus manually. What is more, the speech act that the speaker intended to produce is sometimes

ambiguous and even when having access to a linguistic production: it is not always possible to clearly identify the speaker's intention. This annotation therefore involves an interpretation on the part of the researcher, which is, in part, necessarily subjective.

7) When the sociological information about the speakers in a corpus is known, it is very easy to use this for comparing the productions of different categories of speakers such as men and women, people from different age groups, from different socio-economic backgrounds, etc. These data are typically not part of the corpus itself, but are listed in the metadata (see Chapter 6).

8) The main limitation concerns the availability of diachronic data. Spoken corpora in particular do not go back further than the second half of the 20th Century, which considerably limits the generalizations that can be drawn regarding the evolution of languages, since this primarily takes place in speech. In addition, diachronic corpora only offer a list of certain textual genres, such as literature and legal texts, which also limits the generalization of the conclusions that can be drawn from them.

2.11. Further reading

The usefulness of corpora for studying lexicon, syntax, pragmatics, sociolinguistics and discourse is introduced with English examples in the book by O'Keeffe and MacCarthy (2010). The role of corpora for studying lexicon is also very well illustrated in Szudarski (2017), whereas Rühlemann (2018) offers a similar introduction to pragmatics. Regarding phonology, an interesting discussion on the usefulness of corpora can be found in Cole *et al.* (2011). In the area of morphology, Baayen (2008) discusses how corpora can be used for conducting studies in the area of productivity, a central theme in morphology.

3

How to Use Corpora in Applied Linguistics

In this chapter, we will continue to explore the multiple uses of corpora in the different areas of applied linguistics, in order to complement the presentation of theoretical linguistics in Chapter 2. In particular, we will see how corpus data can be used for studying the language of specific groups such as children, individuals with language impairments and foreign language learners. We will then illustrate the role of corpora as a tool for teaching languages, as well as for creating dictionaries. Finally, we will discuss the uses of corpora outside the language sciences, in order to study literary texts, and also within the legal framework.

3.1. Language acquisition

In order to study language development in children through observation, the use of empirical methods is essential. It is for this reason that this field has been a pioneer in the development and sharing of corpus data. At the beginning of the 20th Century, several researchers systematically studied the language of their children by means of notebooks, where they recorded their observations (e.g. Stern and Stern 1907). These notebooks allowed them to determine the main stages of language acquisition. However, they did not offer a detailed view of the acquisition process itself. In fact, past the stage of the first words and telegraphic speech, which slowly takes place between 1 and 2 years old, typical language acquisition then experiences a strong acceleration between 2 and 4 years old, to such an extent that it is impossible for parents to observe all the changes that take place simultaneously during this period, simply by listening to their child speaking. In order to study these rapid changes, it is necessary to be able to keep track of the exact language produced by the child at different points in their development and to analyze every aspect in detail. These meticulous observations were made possible thanks to the arrival of magnetic tape recorders, which led to the creation of the first real language acquisition corpora in the 1960s, with significant advances for our understanding on the stages of language development. For example, Brown (1973) studied the morphosyntactic development of three children and derived from it an acquisition stages classification, which remains largely valid. Given the crucial importance of data access and the complexity of collecting language acquisition corpora, linguists quickly understood the interest of sharing their resources. This led to the creation of the CHILDES database (MacWhinney and Snow 1985) which offers free access to numerous corpora in different languages (see Chapter 5, section 5.5).

Language acquisition corpora show some specificities when compared to other corpora. First, children's language develops mainly during the preschool years. The development of spoken language largely precedes the start of the written language learning process. This is why language acquisition corpora are by nature spoken corpora, which require a written transcript in order to be analyzed. The transcription itself poses certain challenges, since children who acquire language produce mistakes (see Chapter 7 for a discussion regarding the different ways of annotating such mistakes). Studying these mistakes provides valuable clues for understanding the acquisition process. It is therefore advisable not to erase them in the transcriptions, but to keep a log of them so that they can be easily identified (e.g. using categories such as "consonant substitution", or "truncated word"). Finally, in the recordings, children interact within their environment (often at home), with people they know well, most often, their parents. For this reason, language acquisition corpora frequently include language samples produced by children as well as by adults. This configuration makes it possible to analyze the way in which adults respond to children's language, as well as the connections between the language that children hear from their parents and their own productions. This information represents valuable clues for studying acquisition mechanisms, and these are particularly valuable for theoretical frameworks which attribute a key role to social interactions as the source of language acquisition (e.g. Tomasello 2003). Finally, the last characteristic of children's corpora is that they are relatively homogeneous, unlike the corpora containing language produced by adults, since children do not master different language registers. These corpora are therefore easier to compare than the corpora produced by adult speakers.

Two types of corpora can be considered when studying children's language. On the one hand, we have so-called *longitudinal* corpora, in which one or more children are regularly recorded over a period of several years, and on the other hand, we have *cross-sectional* corpora, in which groups of children of different ages are recorded only once. Each corpus type has its own advantages and disadvantages. The great advantage of longitudinal corpora is that they make it possible to study the evolution of language much more precisely than cross-sectional corpora, since these contain samples collected at close intervals. Their disadvantage is that they include samples from a very limited number of children. Given the importance of individual differences in language acquisition (Kidd *et al.* 2017), these studies can sometimes pose problems for the generalization of results. Certain environmental factors in particular such as the socio-economic level of the families (Hoff-Ginsberg 1991), as well as the child's gender (Karrass *et al.* 2002), have been shown to influence the speed of language acquisition and should be taken into account. Conversely, cross-sectional corpora provide data for larger groups of children, which makes observations easier to generalize. On the other hand, cross-sectional corpora impose many limitations for studying the evolution of language in detail, because children in the different groups have often at least a year difference.

The main limitation to the use of corpora (whether longitudinal or cross-sectional) for studying children's language is that they contain only a small portion of the language that children produce at a given time. Typically, a longitudinal corpus includes a recording of a few hours, gathered every one to three weeks. Tomasello and Stahl (2004) calculated that this data collection rate only makes it possible to capture between 1 and 1.5% of the language produced by the child. This percentage is enough for studying the development of frequent linguistic phenomena such as the acquisition of verbal morphology. However, it is insufficient for a number of rarer phenomena such as, for example, the acquisition of the passive voice form. More recently, some denser corpora have been developed in order to represent a larger proportion of children's language, thus making it possible to trace the appearance of rarer linguistic phenomena. The most extreme case is the corpus by Roy (Roy *et al.* 2012) who videorecorded his child's language development for 10 hours every day, from birth to the age of 3 years.

A second limitation inherent to the use of corpora is that even very dense corpora can only be used to study language productions. However, it is well known that a comprehension and production of language do not develop in a totally synchronous manner (e.g. Rowland 2013). Corpus studies should therefore be combined with experimental studies in order to provide complete answers regarding language acquisition phenomena. However, corpora are ideally suited not only for answering questions regarding the emergence of various phenomena in children's language production (such as the ability to produce certain words or syntactic structures) but also for quantifying their development (for instance, lexicon size and the number of words in each grammatical category at different stages of language acquisition).

The first study that we present to illustrate the use of corpora for the study of language acquisition concerns the acquisition of the French verbal system for expressing temporal references. Parisse and Morgenstern (2012) studied the development of language in two girls who were recorded between the ages of 1 and 3 years. In this study, the temporal references were simply divided into three categories: *past*, *present* and *future*. The authors retrieved all the occurrences of verbs (eliminating auxiliaries), and on a separate file, the utterance produced, the grammatical form used and the situational context. They also annotated whether the utterance had been used in a context referring to a present, past or future event and whether it implied a temporal shift with respect to the speech point. This annotation carried with it a type of interpretation, insofar as the children's statements were not made with correct and full-fledged linguistic forms, especially at the beginning. In order to categorize children's sentences, the authors followed their own understanding of the situation, the parents' reactions, as well as their reformulation of their children's utterances. The results showed that the two children started by producing constructions in the present and in the imperative, followed by constructions in the past perfect tense (*passé composé*) and in the periphrastic future (*je vais aller*). It was not until later, at the age of 2 years, that children began to produce constructions using the past perfect – *imparfait* – (*j'allais*) and the future (*j'irai*). However, thanks to their annotation of temporal references, which was not based exclusively on the grammatical form produced but on the communicative intention of the child, the authors were able to observe that children refer to future and past events long before they are able to use the correct verbal forms in order to do so. It would seem that children acquire the semantic functions associated with time before they can develop the grammatical means for expressing them. This study not only illustrates how longitudinal data can bring to light the progressive development of a language aspect in children but also stresses the importance of data annotation and the valuable information that can be gathered when taking into account the language shared between children and their parents.

Parisse and Le Normand (2000) also studied the way in which morphosyntax develops in children who are learning French as a mother tongue. Their study was based on a cross-sectional corpus that included recordings from 27 two-year-old children. This corpus has been fully tagged into morphosyntactic categories using an automatic part-of-speech tagger with the aim of comparing the types of sentences produced by children and those used by adults. In order to have a suitable point of comparison for adult language, a corpus including the language produced by adults while interacting with children was gathered from the CHILDES database. Results indicated that 2-year-old children use remarkably similar sentences to those produced by adults, at least in terms of the frequency of each morphosyntactic category found in their speech. A comparison of the most frequent sequences of two and three morphosyntactic tags in the two corpora also revealed a great similarity between adults and children. Although adults produce many more different sequences, which reflects the greater complexity of their language, a large number of sequences produced by children were also found in adults. Finally, the authors observed that when children produced utterances longer than one word, they made more syntactic errors than adults, a finding that can easily be explained by the fact that children's language is still developing at the age of 2 years. The key point of this study is that it made it possible to bring to light the connections between the morphosyntactic sequences produced by adults and children, and this, from as early as the age of 2 years, when children start producing sentences longer than one word. From a methodological perspective, this study also showed that part-of-speech taggers can be successfully used on children's speech, because the analysis of their utterances did not generate more tagging mistakes than those produced by adults.

3.2. Language impairments

In the same way as corpora make it possible to study language acquisition in children experiencing typical development, they also provide important information about language acquisition in atypical populations such as children with autism spectrum disorder or specific language impairment. The main advantage of using corpora to study such populations is that they provide an overview of the communicative strategies they deploy in order to compensate for certain language and communication difficulties. For example, corpus data make it possible to analyze the way in which a child with limited syntactic skills manages to ask for an object or to ask a question. These also provide an overview of a child's linguistic competence when they find themselves in a familiar situation, offering them better chances of displaying all of their skills. This feature becomes of crucial importance when studying the speech of children with autism spectrum disorders, who suffer from a high level of anxiety when facing unknown situations. Finally, corpora are also useful for comparing the skills of children and patients in different and natural discursive situations (Da Silva Genest and Masson 2019).

Corpora are becoming increasingly used for identifying linguistic markers charactering certain language disorders in adult populations, for example Alzheimer's disease (Fraser and Hirst 2016), or in patients suffering from various forms of aphasia (Wright 2011). They also make it possible to study the way in which patients suffering from schizophrenia communicate (Howes *et al.* 2017). Some researchers are even trying to develop automatic tools to help diagnose language disorders by analyzing the use of these typical markers evidenced in corpus data (Bull *et al.* 2016).

The main limitation to the use of corpora for studying atypical populations (which is also valid for the study of normally developing children) is that corpora only provide information on spontaneous linguistic productions. However, when a certain linguistic phenomenon is not found in the corpus, it does not necessarily mean that the child or patient recorded in the corpus does not have the competence to produce such types of words or sentences. It only means that they did not employ them during the recorded sessions, either because they did not have the opportunity to produce such an element, or because they deployed an avoidance strategy due to the fact that they could not master such constructions. As we have already seen, this limitation is particularly acute for the study of rare linguistic phenomena. As we will later explore in the case studies, when studying children with atypical development, it may be advisable to analyze a corpus of children with typical development (or suffering from a different pathology which does not involve the same linguistic deficits) in parallel, so as to favor the comparison between different linguistic productions in similar contexts. In the same way, when it comes to studying the language of adult patients, it is necessary to compare it with the language used by healthy adults, produced in similar contexts.

Another limitation to the use of corpora is that the latter do not make it possible to measure the connections between verbal productions and other cognitive skills, such as working memory or non-verbal intelligence. Gathering data about these non-linguistic skills is often important in order to understand the nature and causes of the language deficits observed. It is for this reason that the study of language production in patients is often carried out via constrained production tasks, such as the ability to name images or to repeat non-words or sentences (see, for example, Seiger-Gardner and Almodovar 2017), rather than based on corpus data alone. Performed in an experimental context, these tasks make it possible to control many linguistic parameters that influence production, as well as to limit the impact of avoidance strategies and to test the existence of connections with other individual differences in working memory or non-verbal intelligence.

The first study that we will discuss in this section compared the linguistic productions of six children with autism spectrum disorder (ASD) with those of six children with Down syndrome. The aim was to determine whether the syntactic development of children with ASD followed a different trajectory compared to another population who also suffers from language impairments. Tager-Flusberg *et al.* (1990) recorded these children twice a month during spontaneous interactions with their mother over a period of 12–26 months. At the start of the recordings, the chronological age of the children ranged from 3 to 7 years. Children were matched between the two groups based on their mean length of utterance[1]. The children chosen for the study came from families with a comparable socio-economic level. Since the children in both groups frequently produced echolalia, for example, by repeating previously heard utterances or by repeating their own utterances several times, the authors compiled a 100-utterance sample per visit for their analysis. On the basis of this corpus, the authors then measured for each

1 See Rice (2010) for a definition and an application of this concept.

child the evolution of the mean length of utterance, the development of the syntactic complexity of their sentences, the lexical diversity and the proportion of each morphosyntactic category. The results showed that children with ASD and Down syndrome have quite a similar development from the point of view of all the measurements made, both at the syntactic and the lexical level. This development also corresponds to the development of typical children, as described in the literature, but taking place with a certain delay. This study revealed that verbal children with ASD do not have a specific deficit regarding the acquisition of morphosyntax (at least from the point of view of production), but only present a delayed development when compared to normally developing children.

The second study concerns the production of noun phrases by French-speaking children with specific language impairment (SLI). Royle and Stine (2013) recorded eight children diagnosed with SLI aged between 5 and 6 years, eight children matched on chronological age and eight children matched on their mean length of utterance (MLU). The two latter groups, called control groups, were used for comparing the productions of children with SLI with those of comparable children (comparable in terms of age or level of linguistic development). Since the linguistic development of children with SLI is delayed compared to that of normally developing children, the group matched on MLU made it possible to establish a comparison with children whose language is at a similar development stage, which is not the case for children of the same chronological age. This comparison enabled the authors to measure the gap between the two groups from the point of view of their linguistic development. Results indicated that children with SLI produce more simple noun phrases (without a modifier) compared to the other two groups. They also tend to produce more omission mistakes when an element is mandatory or to produce substitution mistakes between elements. In addition, children with SLI do not make the same types of errors as normally developing children. While normally developing children mainly produce lexical errors such as the use of an incorrect plural (*les chevals*, instead of *les chevaux*), children with SLI mainly produce phonological errors such as false cuts due to *liaisons* (*le néléphant*) or a lack of elision (*la image* instead of *l'image*). These results add credit to theoretical models that consider SLI to be primarily an impairment affecting the functional aspects of language, such as phonology, morphology and syntax, rather than the lexicon or pragmatics.

3.3. Second language acquisition

The field of second language acquisition is one of those in which the use of corpora has grown exponentially in recent decades. The creation of numerous learner corpora, as well as the development of new methods and annotation tools, has largely contributed to this evolution. While linguists working on the question of second language acquisition have long used learners' productions as a source to build their theories, these data were limited to very small samples or even to single-person studies. Therefore, the generalization potential linked to these data was highly questionable. This led to the creation of real learner corpora, aiming to provide representative samples of this population.

There are different types of corpora containing language produced by learners. The first learner corpora produced at the end of the 1980s were limited to written productions, and these corpora are still the most numerous today (Gilquin 2015), but spoken corpora have also developed recently. Learner corpora often include language produced when performing various kinds of tasks such as essay writing, conversations and descriptions of images. As a result, the status of these productions as samples of natural linguistic productions can be called into question. In fact, classroom writing is an artificial genre, not entirely representative of the way in which students write freely. Similarly, image description is justified for educational purposes but does not correspond to real situations involving spontaneous speech. However, Granger (2008) argues that learner corpora represent "quasi-natural" language, because the situations they represent are typical (albeit not of everyday life situations) of linguistic activities carried out in a language learning situation.

As is the case with other types of corpora, learner corpora have mainly been compiled in English, but there are also resources for other frequently taught languages such as French (see Chapter 5, section 5.5). Some corpora contain language samples produced by learners of different mother tongues, and this information can be found in the metadata (see Chapter 6, section 6.4). These corpora are valuable tools for measuring the role of different mother tongues in the process of acquiring the same foreign language, as we will see below.

Finally, most learner corpora are cross-sectional corpora, including one sample per participant and representing a given moment during the acquisition process, since most of the time learners included in a corpus have a homogeneous level of competence in the foreign language. These corpora are very useful for determining learner competence at a certain level, but considered individually, they do not make it possible to study different acquisition stages. For this, longitudinal corpora are necessary, but these are rare due to the difficulty of sampling the same learners across several years. One way to get around this problem is to compare several cross-sectional corpora including learners from different levels of competence.

The first study we will discuss focused on whether learners at a very advanced proficiency level keep on improving, which would justify the need to define different development stages for learners beyond the so-called advanced stage of acquisition. To do this, in the spoken corpus *InterFra*, Forsberg Lundell *et al.* (2014) defined three groups of French non-native speakers whose mother tongue was Swedish. Each group included 10 speakers. The first group was made up of speakers aged 19 to 34 years who had lived for one to two years in France. The second group included speakers aged 25 to 30 who had lived between 5 and 15 years in France, and the third group included speakers aged from 45 to 60 years who had lived between 15 and 30 years in France. These groups were compared to two groups of 10 native speakers each, who were between 15 and 30 years old and 45 and 60 years old respectively, chosen to match the ages of learners.

The groups of learners were then compared on the basis of five linguistic indicators:

– the number of non-native morphosyntactic forms produced, for example, gender or plural agreement mistakes;

– the number of left dislocations, in sentences such as "*moi, si tu me demandes, il a tort*" (literally: me, if you ask, he is wrong), which are typical of spoken French;

– the number of formulaic sequences such as collocations;

– lexical richness, calculated based on the number of words with high, middle and low frequency in corpus data that are used;

– fluency, measured by articulation speed and utterance length between two pauses.

The results indicated that the group of learners with 5–15 years of residence differed from the 1–2 years of residence group on the following criteria: use of formulaic sequences, lexical richness and fluency. The group having lived longer than 15 years in France did not differ from the 5–15 years of residence group on any of these indicators. However, speakers who had lived the longest in France managed to pass for natives in a listening discrimination test administered to French native speakers, unlike the speakers from the 5–15 years of residence group. This indicates that some form of progression must have taken place between these two groups, but which could not be measured through the tests chosen for this study. Furthermore, all the groups of learners differed from native speakers (but did not differ from each other) on the assessment of morphosyntax, which seems to indicate that this is an area which can remain beyond the reach of even the most advanced learners. This study thus showed that language continues to develop beyond the so-called advanced acquisition stage and that this progression is not uniform among the different dimensions of language. As the lexicon continues to progress, certain aspects of the language system such as morphosyntax remain at a non-native level, even at truly advanced acquisition stages.

The second study that we will introduce compared the use of two English discourse markers, *in fact* and *actually*, by learners of two different mother tongues and by native speakers. Buysse (2020) compared the oral productions of French-speaking and Dutch-speaking learners of English in the *LINDSEI* corpus with the productions of native English speakers in the *LOCNEC* corpus, a corpus which was compiled to be comparable with the *LINSDEI* (see Chapter 4 for a definition of the concept of comparability). The interest in comparing French and Dutch speakers is that there are different translation equivalents for the English markers in both languages. While Dutch has two markers which closely resemble those in English (*eigenlijk* for *actually* and *in feite* for *in fact*), French only has one close marker, which is *en fait*. So, in French, *actually* has no translation equivalent of its own.

In her study, the author first performed a frequency analysis regarding these two markers in the three sub-corpora. The results indicated that *actually* is significantly more common than *in fact* among Dutch native speakers in comparison to French speakers. Conversely, *in fact* is significantly more common than *actually* in the speech of French speakers compared to Dutch speakers. On the basis of the literature, she then identified all the possible functions for these markers in English, such as introducing an elaboration or a contrast, and then annotated all the occurrences of these markers according to one of these functions. This analysis allowed her to show that learners use all the possible functions that the markers offer, even if their respective frequency varies a little between French speakers and the other two groups. That being said, the low number of occurrences of the marker *actually* among French speakers (56 in all) prevents a quantitative analysis of the differences between its different functions. In summary, this study demonstrated the influence of speakers' mother tongue on the use of discourse markers in a foreign language, and in particular, the importance of having a similar marker in L1 to help learners use markers appropriately in a foreign language. Indeed, Dutch speakers, who have two very similar markers in their mother tongue, use *in fact* and *actually* in the same way and in the same proportions as natives. On the other hand, French speakers tend to under-use the marker, which has no direct equivalent in their mother tongue (*actually*) and to overuse the other marker (*in fact*), to perform the same functions, as they would do in French. This study thus indicates that negative transfer effects occur even among advanced learners.

3.4. Language teaching

In addition to collecting learner corpora as we discussed earlier, the area of language teaching currently makes an extensive use of corpora produced by native speakers (see, for example, Sinclair (2004) and Cheng (2010) for literature reviews). Corpora including different genres are used for the preparation of teaching materials, in order to present learners with real-life communication examples. Corpora also help to set these examples in a much richer context than traditional dictionaries and grammar textbooks. Finally, using frequently updated corpora makes it possible to provide examples of usage that better match the reality of contemporary speakers than conventional tools, whose examples are aging rapidly and which often represent only a normative usage that is often disconnected from the reality of native speakers.

In the field of vocabulary in particular, the use of corpora makes it possible to empirically provide lists of the most frequent words in a certain field, which should therefore be taught as a priority. Another key point for mastering a foreign language is to know, apart from the meaning of isolated words, certain elements of phraseology, in other words the typical linguistic sequences in which a word occurs, for example for the word "*knowledge*", "*to acquire knowledge*", "*knowledge gain*" or "*prior knowledge*". Some researchers even think that these elements should be taught as lexical units (Kennedy 2003). On this point, corpora have become very useful resources, because they make it possible to automatically retrieve the most common phraseological elements for a given word.

In addition, corpora provide examples of spoken language, which are clearly more realistic than the constructed dialogues contained in most language methods. Given that they include natural interactions, corpora include reformulations, hesitation markers, turn-taking devices, etc., which are not reproduced in artificial dialogues, but which are important for learners to master since they are an integral part of language uses among native speakers. Finally, the creation of learner corpora has also made it possible to bring a new dimension to language teaching, by allowing learners to consult non-native productions and to compare them with native productions. Access to such data enables learners to become aware of the differences between their productions and those of natives. In addition, these corpora often contain an annotation of errors, which favors an explicit learning process and enables learners to become conscious of typical errors and to avoid them.

An important question for language teaching is to determine to what extent the corpora developed for linguistic research can be reused as such in the classroom. On the one hand, there are many advantages to letting learners use corpora by themselves, for example, by teaching them to search for word occurrences using a concordancer. This practice induces active reflection on the language, when it comes to determining what to look for and how to look for it, which enhances learner autonomy and stimulates students to become involved in the learning process (Bernardini 2004). On the other hand, the use of corpora built for research purposes implies both the need to learn how to use corpus searching tools, as well as the ability to interpret the numerous concordance lines retrieved by such operation, and in particular, how to sort relevant occurrences from noise. In certain teaching contexts, these barriers prevent the use of corpora.

Furthermore, Braun (2005) argues that several features of corpora have been created for research purposes and renders them unsuitable for classroom use. While it is true that their large size is essential for answering many research questions in linguistics, this makes them both impractical and of little use for learners. For the latter users, in fact, a more limited number of well-chosen illustrations is better than hundreds of concordance lines. In addition, for corpora to be useful for learners, they should contain annotations of many linguistic phenomena such as syntactic structure and speech acts, in order to make them suitable for later search. In research corpora, these phenomena are still not frequently annotated at a large scale. Conversely, the sets of tags used by part-of-speech taggers are often too detailed to be understandable by learners (see Chapter 7). For all these reasons, smaller and more specific corpora are produced by teachers to better meet the needs of their class. Aston (2001) suggests that learners should start by using these small corpora specifically built for classroom needs (see Reppen 2010a) before moving on to larger, more general corpora when they reach a more advanced learning stage.

Limitations on the use of corpora created for research also apply to the use of raw corpus data for creating language methods. In order to base a language method on corpus data, it is imperative that the corpus chosen is adapted to the target audience, in particular from the point of view of the variety of the language represented, discourse genres, the age of the speakers, etc. According to McCarten (2010), frequency data drawn from corpora should not always be implemented as such in language methods. Indeed, certain infrequent words in corpora are still part of the basic vocabulary of a language. A case in point are the words for naming the days of the week. Conversely, some frequent words in corpora, such as prepositions, sometimes involve concepts or linguistic structures that are too complex to be included at a beginner level. The designers of a language method strike a balance between the frequency information provided by corpora, the perceived usefulness of the word and its learning difficulty. In the same way, trying to include raw spoken data in teaching materials can be troublesome. First, real conversations are often too long to be studied in their entirety, and cutting them poses consistency problems. Second, spontaneous conversations sometimes refer to topics that are uninteresting, inappropriate or simply difficult to understand without the conversational context. For all these reasons, spoken data should often be prepared (choice of theme, predetermined length, etc.) in order to avoid these problems.

In this section, we will introduce two studies which show the usefulness of corpora for language teaching. Each of them compared corpus data with the presentation of the same phenomenon using different language methods. Biber and Reppen (2002) looked at three aspects of English grammar teaching in six widely used language methods for different levels. They identified three elements in these methods:

– the grammar points discussed;

– the order in which they were presented;

– the vocabulary used in the examples to introduce such points.

Then, they compared the examples with frequency data drawn from a 20 million word corpus, corresponding to four different language registers.

In each of the three areas studied, significant differences were observed between corpus data and the presentation of the same phenomenon in language methods. For example, in the section introducing the forms that noun phrases may take in English, most of the methods only indicate a pre-nominal modifiers can be an adjective (*a nice man*), a present participle (*an exciting game*) or a past participle (*stolen goods*). However, in written corpora, nouns are also common modifiers of other nouns (e.g. *metal seat* and *tomato sauce*) and the relationships they express are diverse and complex. This syntactic pattern should also be included in language methods. In addition, the order of presentation for the different grammatical features does not correspond to the uses observed in corpora, especially in the case of verbal tenses. Most methods strongly emphasize progressive forms and represent them as the default form in conversations. However, corpus analysis shows that the most frequent case in many language registers is, on the contrary, the simple aspect. Finally, the authors observed that the verbs used for illustrating different grammatical properties in language methods are not necessarily the most common verbs in real-life language. Although introducing less frequent verbs may be useful for broadening learner vocabulary, it is nevertheless surprising that the most common words are not used, at least for beginner learners. This study showed that the intuitions of language method designers often do not reflect actual language uses. Corpus data make it possible to produce better-suited educational materials to match the realities encountered by learners.

Racine and Detey (2017) also compared the information given in language methods with corpus data, focusing on the question of *liaisons* in French. Producing liaisons is a particularly difficult aspect of spoken French for learners. In fact, producing liaisons correctly requires mastering production constraints (such as identifying the consonant involved in the liaison, taking into account possible modifications to the phonological environment, etc.), as well as the syntactic environment making the liaison either required, optional or forbidden. Most methods of French as a foreign language focus only on the compulsory, optional or forbidden nature of the liaison, which they present in a normative and simplified manner (Racine 2014). However, recent corpus studies have revealed that learners have many problems in producing liaisons, which are little addressed in language methods, or not addressed at all (Chevrot *et al.* 2013). Conversely, corpus studies with native speakers indicate that the contexts in which liaisons are produced may vary from speaker to speaker, different language registers and even ages (see Chapter 2, section 2.1), and contradict the normative data introduced in language methods. Indeed, when learner language is compared to that of native speakers in contemporary corpora, no difference in the production of compulsory liaisons can be observed between the groups, because the standard represented in language methods often does not correspond to the reality of the productions of native speakers. The study thus highlighted the importance of using native speaker corpora in order to compare their production with that of learners. These comparisons may help to better identify the elements to be taught.

3.5. Lexicography

Writing a dictionary requires the use of textual data in order to identify the words that should be included and to illustrate their contexts of use. Since the beginnings of dictionaries, lexicographers have manually collected examples from various sources, mainly literary ones. This focus on literary texts is particularly visible in the case of French, a language for which lexicographers have focused on a formal register of the language. For example, the *Trésor de la Langue Française* has drawn its 430,000 examples from "two centuries of French literary productions" (Pierrel *et al.* 2004). In other languages such as English, however, the use of other genres like journalistic texts is much more widespread.

Regardless of the stylistic genre targeted, the use of quantitative methods linked to corpus linguistics has led to many advances in lexicography and has been one of its main application areas. Indeed, searching for occurrences of words corpora rather than registering them manually over readings has permitted lexicographers to list examples much more easily and to use the frequency information provided by the corpus data, in order to decide which meanings to include and the order in which to present them in the dictionary entry.

Since the 1970s, the first dictionaries making use of corpus data became available in English. The first large-scale lexicographic project involving the massive use of corpus data was the *COBUILD* dictionary, dating from 1987. Following this project, many lexicographers quickly decided to use corpus data by insisting that word census and meaning classification based on a qualitative approach were unreliable (Sampson 2001). However, the use of corpora remained controversial for a long time among lexicographers even in the English-speaking world, in particular as regards the insertion of real examples from corpora into dictionaries, because they were considered unclear and rather unsuitable for unambiguously illustrating word meaning (Hausmann and Gorbahn 1989). This is why, for several decades, corpus-based dictionaries and dictionaries based on artificially built examples have rubbed shoulders on the market. In the English-speaking world, the debate on the advisability of using corpus data is now closed and most of the large publishing houses have at their disposal large corpora on which their dictionaries are based.

This paradigm shift is taking longer to settle in the French-speaking world. Siepmann (2015) quoted a page from the *Petit Robert* website where it was reported that this publishing house still used a manual search approach at that moment for locating occurrences based on reading. Although this web page has now been deleted, still in 2019, there seemed to be no indication on the *Petit Robert* website showing that such a practice had changed in recent years. However, a qualitative analysis of isolated examples comes across many limitations, because it leads to the inclusion of word senses and examples that only partially reflects their real uses and offers a biased vision of the most frequent meanings (see Hanks (2012) for an in-depth discussion of the limitations associated with the qualitative approach in lexicography). In this section, we will show that corpus data make it possible to overcome some of these limitations.

To be sure, corpora offer many uses for lexicography (see, in particular, Walter (2010) for a more in-depth discussion). They make it possible to have an objective vision of the frequency of a word in the language, as well as a census of its different meanings with a high degree of exhaustiveness. They also offer the possibility of determining which meanings are specific to certain textual and discursive genres. The use of very large corpora drawn from the Internet also makes it possible to identify new words and new meanings very quickly as soon as they enter the language, as well as many words which are currently absent from dictionaries. As we saw in Chapter 2, based on the *Google books* corpus, Michel *et al.* (2011) estimated that almost 50% of the lexicon currently in use in English is not listed in any dictionary. Finally, corpora are extremely effective tools for identifying typical collocations between words.

The first study that we will introduce for illustrative purposes stresses the importance of using a corpus adapted to the target audience of a dictionary, rather than a general language corpus. As we said above, in the English-speaking lexicographic tradition, the use of corpora is now the norm, but reference corpora most often refer to the same stylistic genres. Wild *et al.* (2012) studied the value of basing a children's dictionary on a corpus of texts intended for children, rather than on a general English corpus. To do this, they created the *Oxford Children Corpus*, containing over 30 million words drawn from works written for children aged between 5 and 14. Furthermore, they tagged every text fragment inserted in the corpus as belonging to three different levels (5–7 years, 7–11 years, 11–14 years), according to a text classification in force in the education system of the United Kingdom.

The authors then retrieved certain keywords from the *Oxford Children Corpus* in order to compare them with those in the *Oxford English Corpus*, containing texts intended for adults. These lists were retrieved automatically, then manually compared by sorting the keywords into thematic groups for each textual genre. This analysis indicated that the two corpora had different themes. While the authors who write children's fiction talk more about nature and the physical world (including body parts, buildings, objects and time), the authors who write adult fiction mainly deal with politics, religion, work, education, human relations and death. From the point of view of functional words, texts intended for children in the corpus mainly refer to the question of space, whereas the texts intended for adults focus primarily on the temporal dimension. Many differences between corpora are also present in non-fiction genres. In addition, this comparison reveals differences in the way of addressing readers, since children's writers have a more direct and informal style.

The authors argued that these differences should be implemented in children's dictionaries in several ways. First, the differences in themes should lead to a choice of words included in the dictionary based on books written for children, rather than producing dictionaries for children that are only abridged versions of adult dictionaries. In particular, the words chosen should reflect their frequency in the books designed for the age group targeted by the dictionary. A comparative analysis of collocations in the two corpora also indicated that words are used differently in children's literature and that these differences should be reflected in dictionaries. For example, the English word *play* is frequently used with musical instruments and sports nouns in the two corpora.

By contrast, it is more rarely found in collocation with words indicating a role, as in *he played an important role in his failure*, in texts destined for children than in texts for adults. Finally, the authors argued that corpora intended for children provide excellent sources of examples, because they use words in a context which is familiar to them and involve books that they partly recognize and often appreciate.

Siepmann (2015) also argued in favor of the use of specific corpora and more precisely, spoken corpora, as a basis for the creation of dictionaries, in order to adequately account for the meaning of words in colloquial registers. Although the latter are generally included in dictionaries to some extent, their treatment in French dictionaries does not correspond to their actual use in the language. To illustrate this problem, the author used the *Corpus de référence du français contemporain*, a corpus of 310 million words representing written and spoken modes, with a sub-section of 30 million words of informal speaking. The author then extracted from this corpus a list of words tagged as *colloquial*, or *vulgar* in the *Nouveau Petit Robert*.

For all the words studied, the results indicated significant differences between the results of the corpus study and the entries of different French dictionaries. For example, one of the words analyzed was *mec* (guy). The corpus analysis revealed four types of use for this word:

– a man as opposed to a woman (1);

– any male individual (2);

– a boyfriend (3);

– a virile man (4).

However, most dictionaries covered only a part of these meanings. On the other hand, when a meaning was included in the dictionary, it was often done incorrectly. For example, for meaning (3), the *Nouveau Petit Robert* mentioned "a woman's companion", whereas the corpus showed that this use also extends to homosexual couples. Likewise, the *Trésor de la Langue Française* indicated that meaning (2) of the word *mec* listed above is often used in a derogatory way, as in certain collocations like *pauvre mec* or *petit mec* (a poor guy). However, corpus analysis has revealed that the predominant collocations are rather neutral, as in *jeune mec* (a young guy) or positive, as in *mec bien* (a good guy), *beau mec* (a handsome guy). Dictionary examples do not properly reflect the most frequent connotation of the word *mec* in the sense of the male individual. This study showed that the use of spoken corpora makes it possible to offer a more complete and appropriate treatment of frequently spoken words which are hardly ever (or not even) included in dictionaries.

3.6. Stylistics

The stylistic study of literary texts is traditionally based on a qualitative analysis of chosen text excerpts. However, the choice of excerpts is often complex, since the identification of interesting characteristics to be studied in a text is not always visible at a first reading. This is why literary scholars working in stylistics have started to acknowledge the interest of incorporating quantitative methods from corpus linguistics into the analysis of literary texts, so as to have a more objective starting point for identifying interesting themes and relevant excerpts, which should be the subject of a book's qualitative analysis. As we will see below, in the field of stylistics, corpus linguistics tools do not seek to replace qualitative analysis, but only aim to guide the choice of themes and excerpts to analyze.

Another advantage of the quantitative analysis of literary texts is that it provides clues for identifying the author of a text when the latter is unknown or controversial. In this case, a series of linguistic indicators are used for measuring the formal similarities between this text and other reference works by different possible authors. The problem then is to find out the best indicators for identifying the similarities between an author's text and its differences with the texts of other authors and to later apply these measures to the controversial text in order to determine which author this text shares more similarities with. These indicators are, for example, words and sentence length, the list of frequently used words, vocabulary richness, frequent collocations, the position of words in sentences and frequent syntactic structures (Oakes 2009). The main problem with these measures is that they are more effective for establishing differences between text genres than between authors. The length of words in particular, as well as the type of frequent syntactic structures, seem to vary notably depending on the textual genre considered (Santini 2004). Hence, Baayen *et al.* (1996) insisted on the fact that a comparison between authors can only be done on texts belonging to the same genre, because for different genres, the differences between genres prevail over differences between authors. Another limitation to this type of analysis is that the style of the same author varies over time and that some authors mix different styles, depending on the type of characters they represent (DeForest and Johnson 2001). Finally, style also varies according to the author's sex (Koppel *et al.* 2002), and comparisons should therefore be limited to authors of the same sex. For all these reasons, the attribution of a text to an author through this type of analysis cannot provide a definite answer but only an estimate of the probability that a text has been written by a certain author, making it possible (or not) to strengthen assumptions based on other criteria.

In order to illustrate the role of quantitative methods in corpus linguistics for the stylistic analysis of literary texts, we will first introduce a study on theme and keyword identification in the novel *Casino Royale* by Ian Fleming. For this study, Mahleberg and McIntyre (2011) extracted a list of keywords from the novel using *WMatrix* software. Keywords in a text can be generated by comparing the text to a reference corpus (see Chapter 5, section 5.7.2). The authors chose the literary section of the *British National Corpus* (BNC) as a reference corpus. They listed the 150 most common keywords in *Casino Royale* and classified them manually.

In addition to this keyword analysis, the authors classified the semantic areas of the novel, which *WMatrix* software can perform automatically. By combining these two methods, the authors were able to identify important words from the novel. Then, they analyzed the occurrences of these words one by one using the concordance file (see Chapter 5, section 5.7). This analysis made it possible to show how the different uses of these words contribute to build upon the themes of the novel, in a much more detailed way than through the analysis of selected excerpts, because all the occurrences involved in the development of a theme could be identified and analyzed. This study revealed the importance of supplementing the analysis of excerpts with automatic analyses covering the whole of the novel.

The second example that we will discuss in this section deals with the analysis of *blockbusters*, big budget movie film scripts, as a fictional genre. McIntyre and Walker (2010) built a corpus of 200,000 words from blockbuster film scripts. Literary critics of this cinematographic genre noted that male and female characters are given unequal treatment, with a clear bias in favor of men. Gender distinctions are also reflected in the stereotypical roles occupied by the different characters. In particular, physical prowess is a characteristic trait of the male hero, and most of the time, the latter operates at the margins of society, whose established power he rejects. The authors wanted to verify whether the characteristics identified by literary critics were reflected in the language of film scripts.

The results indicated that male characters have a much longer speaking time than the female characters, more than four times more words in the corpus. Likewise, the topics men and women talk about (identified from the keywords of the corpus) also differ. Words linked to power were dominant in men's discourse, which again confirms one of the gender stereotypes. On the other hand, the analysis did not show a tendency to reject authority on the part of male protagonists, as illustrated by the frequent use of terms of address such as *sir*. Nevertheless, the authors added that a qualitative analysis would be essential to learn more about this last point. In short, this study confirmed that corpus analysis can help us to study not only the style of an author or a text genre but also the way in which the different characters in a written production are represented.

3.7. Legal linguistics

In recent decades, the expertise of linguists has been increasingly sought in the context of legal cases, for questions relating to all language areas. For example, linguists are sometimes required to authenticate a person's voice on a recording or to determine their geographic origin. Linguists also analyze complaints related to the linguistic complexity of certain public information documents, when users argue that they have misused a product or misunderstood their rights due to such excessive difficulty (Levi 1993). Other disputes relate to the meaning of a certain word in everyday language in relation to its specific legal uses, in order to determine whether it is reasonable to expect the public to know its specific use. For example, Sinclair (quoted by Coulthard *et al.* 2017, pp. 110–111) had to determine whether the legal meaning associated with the noun *visa* (which implies the right to request permission to enter or to leave a country, rather than the right to enter it directly) was the meaning of this word in everyday language. By means of a corpus study of *The Times* newspaper, he analyzed the 74 occurrences of this word and concluded that most occurrences suggested that a visa gave the right to enter or to leave a country, rather than requesting authorization to enter the country, and that it was unreasonable to expect citizens to know its legal meaning. Finally, some cases relate to the interpretation of a statement. For example, Tiersema (2002) argued that when on cigarette packages, the tobacco industry mentions that they "contain tar", they are violating the Gricean maxim of quantity (Grice 1989) by not providing enough information. The consequences of the presence of this substance for health must be inferred, and this is not automatic for all speakers, because it depends on their knowledge level about the subject.

While some of the questions mentioned above can be answered using theoretical knowledge, as in the case of Gricean maxims, many others require the analysis of corpus data. This is particularly the case of requests concerning the attribution of a text to one or more alleged authors. We have already mentioned the question of textual attribution in the context of literary stylistic analyses. This problem arises somewhat differently in the case of judicial inquiries, in particular because of the type of linguistic material involved. In this context, linguists are most often required to examine threatening or blackmail letters, suicide notes, ransom demands and police statements. However, dealing with this type of material poses a certain number of methodological challenges (Cotterhill 2010). First of all, these are extracts from generally very short texts. They do not exceed a few pages, and in the case of e-mails even less, unlike literary works. This small amount of data considerably complicates their quantitative analysis, to the point of making it impossible. The data that linguists analyze in these situations are often unique texts whose author is unknown and for which it is necessary to identify the gender, age and author's region of origin for profiling procedures. In other cases, several texts exist and several potential authors have already been identified. This favors a more complete analysis using corpus linguistics methodology, by comparing the excerpts to other productions of these different authors and by measuring their similarities. But even in those cases, some problems arise. Often, other productions by the same author come from very different text genres or were produced at a different time from the text that requires identification. As mentioned previously, these factors create many variations in style. Finally, analyses should be carried out during a very short period of time in order to provide elements for the courts to decide whether or not to lay charges. Despite these limitations, analyses of textual productions have made it possible to provide decisive elements in certain cases (see, for example, Kredens and Coulthard 2012).

By means of an example, we will briefly introduce the case of Derek Bentley, a young person convicted of murder and executed in the United Kingdom in 1953 on the basis of his police statement (Coulthard *et al.* 2017). At his trial, Bentley claimed that his police statement did not match what he had declared, as well as incorporated material added by the police officers. Around 30 years after the conviction, the file was reopened by Coulthart and his colleagues, who observed that it was indeed unlikely that certain elements of the declaration could have been produced by Bentley, because certain formulations corresponded to police speaking mannerisms rather than those of a teenager. One of the key points of this analysis focused on the use of the temporal connective *then*. This word appears 11 times in a 582-word statement and, what is more, it is employed in a syntactic position, which does not sound typical of the common spoken language. In the declaration, *then* is positioned after the pronominal subject (*I then*), whereas in the common language, this word is rather used at the beginning of a clause (*then I*). Coulthard *et al.* compiled a comparable corpus from two types of witnesses: on the one hand, statements by civilian witnesses in similar cases to that of Bentley and, on the other hand, statements made by police officers.

The results indicated that the word *then* appeared once every 930 words in witness statements, as opposed to once every 78 words in police statements. Thus, frequency of this word in Bentley's statement, corresponding to one occurrence every 53 words, was much closer to police language rather than to that of the witnesses. A search for this same connective in the COBUILD corpus of spoken English showed a frequency of one occurrence every 500 words, which matches the use by witnesses rather than the police. Even more strikingly, the structure of the type *then I* appeared 10 times more frequently in the COBUILD corpus that the *I then* construction, repeatedly employed in the declaration. This result confirmed that it is very unlikely that Bentley spontaneously could have produced those sentences and that the police certainly added elements to his story. Based on this new analysis, the case was re-evaluated and Bentley was posthumously acquitted in 1998.

3.8. Conclusion

In this chapter, we have shown how corpus linguistics can be of use in different areas of applied linguistics. We have seen that language acquisition corpora have helped us to understand the different stages of this process, in particular regarding the study of the associations between the language that children hear in their environment and their own productions. We have seen that corpus analysis makes it possible to better characterize the language specificities of people suffering from language and communication impairments, by studying the different ways in which these patients interact in a natural environment. We then reviewed the multiple applications of learner corpora to better grasp second language acquisition processes and showed how these corpora can be integrated into teaching materials. We also discussed the increasingly widespread use of corpora as a basis for the creation of dictionaries and showed that these data help us to overcome many inherent limitations of a purely qualitative approach to writing dictionaries. Finally, we discussed the different ways in which the corpus linguistics methodology makes it possible to provide valuable tools for the stylistic analysis of texts, as well as for author identification in a legal framework.

3.9. Revision questions and answer key

3.9.1. *Questions*

1) In addition to morphosyntax discussed in this chapter, what are the other aspects of language acquisition that are well suited for corpus-based research?

2) What are the methodological aspects that should be considered when carrying out a corpus study with children with autism spectrum disorders?

3) What type of learner corpora should be used for tackling the research questions below?

a) Study of the development of negation in French during the process of acquiring French as a foreign language.

b) Study of the role of culture in the ability of learners to produce speech acts of requesting in French.

4) Can we use corpora to teach the pragmatic aspects of language, such as politeness, to learners?

5) What are the advantages and disadvantages of using real examples drawn directly from corpora in dictionaries?

6) Imagine a research question in literary stylistics, which could be appropriate for carrying out a quantitative corpus study.

7) Imagine another situation apart from the ones described in this chapter in which corpus analysis would make it possible to detect language uses condemned by the law.

3.9.2. *Answer key*

1) As is always the case with corpus research, the more easily an element is searchable using surface features of the language like unannotated words and sentences, the easier it is to include in a corpus quantitative analysis using raw data. For example, this type of analysis is quite suitable for studying the vocabulary growth of a young child or, more specifically, the emergence of certain words in their lexicon. If the corpus has been annotated, as is the case of many corpora in the CHILDES database, other analyses become possible. For instance, it is possible to study the type of lexical errors made during various developmental stages or the diversification in the repertoire of speech acts which are available to the child.

2) To begin with, we should bear in mind that the use of corpora for studying the productions of children with atypical development meets the same limitations as for typical children: only the aspects linked to production can be studied and they do not make it possible to ensure that an element which is absent from the corpus actually means an inability to master it. This second point can be particularly problematic in the case of children with atypical development, because they develop compensation strategies that

allow them to avoid the confrontation of elements which they find problematic. In the case of children with ASD, it is important to write down the different indications regarding the linguistic profile of the recorded children and to include them in the metadata, since autism represents a broad spectrum of skills and deficits which could lead to comparing children with very different linguistic and cognitive profiles. Above all, it is fundamental that any child recorded in the corpus has been diagnosed with ASD according to the diagnostic tests recognized in the literature, rather than following the mere indication of the pediatrician. Next, information on cognitive skills (non-verbal IQ, working memory) is also important, ideally at different time frames in the case of longitudinal corpora. Finally, it is important that data collection takes place in a context that is familiar to the child and involves as many familiar activities as possible in order to limit any blockages due to anxiety issues.

3) a) In order to study the development of negation in French during the process of acquiring French as a foreign language, it is advisable to use a longitudinal corpus covering different acquisition stages or, else, various cross-sectional corpora of learners at different levels. Negation takes different forms in spoken and in written French (optional use of "*ne*" in spoken data). This type of study should also compare acquisition processes in spoken and written data.

b) In order to study the role of culture on the ability of learners to produce speech acts of requesting in French, it is advisable to use a corpus comprising learners of various languages and cultures. Comparing spoken and written data could also be useful. In order to determine the causes of possible differences between learners and native speakers, the use of comparable corpora produced by native speakers of French would be advisable.

4) Yes, it is possible to teach the use of politeness through corpora. That being said, as we discussed in Chapter 2, it is difficult to identify certain pragmatic uses automatically by means of a corpus search, because these uses are not transparently linked with the linguistic form used. In many cases, it is necessary to analyze the corpus manually so as to find interesting occurrences. However, automatic corpus analysis provides many examples of politeness routines, as the ones related to the opening of a conversation, to its closure or, to speech acts such as apologizing. This could be achieved by searching for specific locations in the interactions (the first or the last lines of exchanges), or through keywords like *sorry* or *excuse me*.

5) The main advantage of using real examples is that they appropriately match the language as it is used by speakers. These examples also reveal the phraseological constructions into which words are frequently grouped, making it possible to provide a rich illustration of their uses. On the other hand, these examples are often too long to be inserted into dictionaries without making any changes. Furthermore, the actual uses of words do not always make it possible for their meaning to be inferred, and therefore do not necessarily accomplish their illustrative role as intended in dictionaries. For example, one of the occurrences of the word *colline* (hill) in the *Le Monde* corpus (year 2012) is "*Sur la colline, tout le monde est allé voter*" (*On the hill, everyone went voting*). This sentence would be a poor example for understanding the meaning of this word. There are many other similar cases in the corpus. Word occurrences should be carefully sorted in order to keep the examples which are sufficiently concise and which offer an illustration making it possible to infer the meaning of the word.

6) Corpus linguistics methodology makes it possible to make contributions on many aspects of literary stylistics, since the search can be based on a keyword analysis of the book in question. For instance, such an analysis could try to identify the important themes in Molière's different plays, as well as to assess whether there are differences in the way in which male and female characters approach such themes.

7) A very good example of the use of corpus linguistics concerns the problem of plagiarism. Today, plagiarism has become increasingly easier to achieve thanks to the wide availability of texts online. By analyzing very large corpora, this crime has now become more easily identifiable by automatic means. Indeed, on a sentence longer than seven to eight words, the probability of producing exactly the same construction as somebody else – by chance – is close to 0 (Sinclair 1991). Such a sentence can therefore only be an example of plagiarism.

3.10. Further reading

The role of corpora for studying language acquisition in children is described in a very accessible way by Diessel (2009). The use of corpus data to study language impairments is discussed by Ferguson *et al.* (2009). In the area of second language acquisition, the most complete resource concerning learner corpora is Granger *et al.* (2015). Reppen (2010a) and Timmis (2015) offer introductions to the use of corpora in the classroom, and Aijmer (2009) brings together various scientific contributions to this question. A practical introduction to corpus-based lexicography can be found in Atkins and Rundell (2008). For an introduction to corpus stylistics, see Semino and Short (2004) and Mahleberg (2014). Finally, the contributions of linguistics to the legal field are presented in detail by Coulthard *et al.* (2017), and the contribution of corpus linguistics in particular in Cotterill (2010).

4

How to Use Multilingual Corpora

In this chapter, we will discuss the main characteristics of multilingual corpora, as well as their different uses. First, we will discuss the advantages and disadvantages of two types of multilingual corpora, namely comparable corpora and parallel corpora. We will see that one of the great difficulties inherent in the use of comparable corpora is the need to define a neutral term of comparison, called *tertium comparationis*, which enables us to measure similarities and differences between languages. We will discuss the different possible terms of comparison, depending on the type of research question being considered. Parallel corpora make it possible to compare texts in their original language, with the corresponding translation into one or more languages. We will discuss the particularities of translations as a text genre and show that, due to these particularities, they cannot be used as if they were original language texts. In the rest of the chapter, we will illustrate the use of multilingual corpora in the fields of contrastive linguistics, translation and bilingual lexicography.

4.1. Comparable corpora and parallel corpora

Multilingual studies can be based on two types of corpus data. First of all, comparable corpora contain original texts in different languages. These corpora are built so as to make samples as similar as possible between languages, and to prevent comparison bias. For example, it would be inappropriate to compare French editorials with English dispatches, even though these two types of texts belong to the journalistic genre. Indeed, their many differences in communicational aims and content make them different in nature, and such disparities could mask differences between languages. It is necessary to neutralize the differences in the type of data used in order to bring out the differences between languages. According to Johansson (1998), the parameters that need to be controlled in order to compare languages include:

– the time when the texts were written;

– their discursive genre (descriptive, argumentative, etc.);

– the type of audience targeted and their field (law, science, etc.).

For example, in order to study the linguistic differences between French and English, one possibility would be to create a comparable corpus of leading articles from journalistic sources with a similar political orientation, published during the same years.

Parallel corpora containing texts in one or more original languages, and their translations into one or more languages, represent the second type of multilingual corpora. It sometimes happens that parallel corpora contain only texts translated into different languages from another language that has not been included in the corpus, or it may occur that the original text cannot be identified among all the texts. As we will see later, the use of corpora in which source languages and target languages remain unidentified poses major problems for contrastive linguistics, due to the special status of translations as a discursive genre (see also Lauridsen (1996)). It would therefore be advisable to use parallel corpora in cases where source and target languages are clearly identified. These are called directional parallel corpora, which refer to the translation direction of source and target languages. Some parallel corpora are even bi-directional, where all the languages they contain are alternately source and target languages. These corpora are particularly valuable for contrastive analyses, since the equivalences between languages are often different in the two directions of translation (see section 4.4).

Both comparable and parallel corpora have many advantages, and also some disadvantages, which we will discuss in the rest of this section. First of all, we should point out that the use of these two types of corpora is not mutually exclusive. On the contrary, the disadvantages of one type can often be counterbalanced, at least partly, by the advantages of the other, and vice versa. This is why many authors are in favor of carrying out contrastive studies on the basis of both comparable and parallel data, when available corpora and time allow for it. We will see examples of such studies later in this chapter.

The main advantage of comparable corpora is their great simplicity of access. *A priori*, it is possible to create a comparable corpus for any language pair, provided that digitized texts of a comparable nature are available in each language. In the case of languages that have already been the subject of numerous corpora, as is the case for European languages, Aijmer (2008) argues that it is often possible to compile a comparable corpus based on existing monolingual corpora. Another advantage of these corpora is that they only contain language samples originally produced in each of the languages, which guarantees their authenticity when compared to translations.

The major drawback of comparable corpora is that researchers have to find data that are highly similar in different languages, in order to avoid blurring comparisons, as we have already discussed. In addition, usage conventions may vary considerably between languages even when the same text genre exists in both, which makes them difficult to compare.

In addition to the difficulty of identifying suitable corpora, from a linguistic point of view, the main limitation regarding the use of comparable corpora is the need to find a neutral term of comparison, undeformed by the prism of either language. Finally, we should point out that while linguistic features can be identified when comparing words or syntactic structures in different languages, these traits are nonetheless difficult to annotate systematically. Indeed, they require a complex type of linguistic interpretation and analysis on the part of the annotator, which, in many cases, implies that the results of the annotation may differ when performed by several annotators (see Spooren and Degand (2010) for an in-depth discussion of this problem and Cartoni *et al.* (2013a) for an illustration in the field of connectives). In Chapter 7, we will discuss the difficulties associated with manual annotation and possible solutions to improve their reliability.

Unlike comparable corpora, the main advantage of parallel corpora is that they guarantee excellent comparability between languages, since the texts they contain are the same. These corpora make it possible to look for equivalences between words, syntactic structures and discursive phenomena, without having to set points of comparison. As a result, comparing languages through the use of parallel corpora is greatly simplified in contrast to comparable corpora because annotators can keep a track of translation equivalents without having to annotate syntactic or semantic features. This method is called *translation spotting* in the literature (Véronis and Langlais 2000) and can also be carried out, in part, thanks to automatic tools. On the other hand, the main disadvantage of parallel corpora from the point of view of linguistics is that they contain only a small portion of original texts, whereas the rest of the material is made up of translations. As we will see later in section 4.3, using translations as a mirror of linguistic practices can also have its drawbacks.

Another practical problem associated with the use of such corpora is their limited availability. In fact, not all languages or discursive genres are regularly translated. In most cases, translations correspond to written genres, often related to the administrative or the literary field (Mauranen 1999).

In addition, language pairs that are regularly the subject of direct translations from one into the other are also limited. What is more, these corpora often include a single source language and a single target language, which makes it impossible to generalize results beyond that particular language pair.

In order to overcome certain limitations pertaining to parallel corpora, the ideal would be to work with a bi-directional corpus, where both languages are alternately source and target, since these corpora make it possible to combine the two types of multilingual data discussed above (comparable and parallel). Bi-directional corpora offer the possibility of studying equivalences in both translation directions through the use of parallel corpora.

In addition, these corpora can be used as comparable corpora, produced in very similar situations, when analyzing only the original language portions of the corpus, as illustrated in Figure 4.1.

Certain corpora fulfill these conditions, such as the Europarl Corpus, a corpus of debates at the European Parliament, where each member employs their own language and whose exchanges are later transcribed and translated (see Chapter 5 for a list of these corpora).

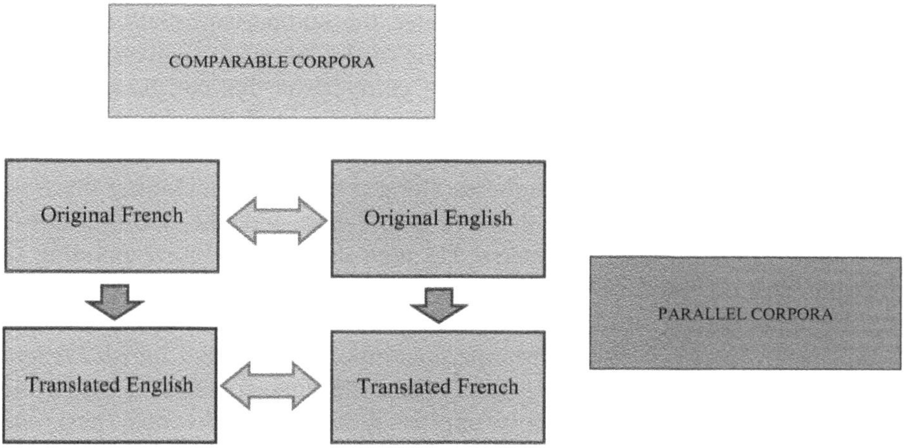

Figure 4.1. *Comparable and parallel corpora that can be retrieved from a bi-directional corpus. For a color version of this figure, see www.iste.co.uk/zufferery/corpus.zip*

4.2. Looking for a *tertium comparationis*

One of the main difficulties inherent in contrastive studies is to find a suitable point of comparison between languages. The problem is that, by nature, comparing two languages implies comparing systems that are partly incommensurable. Therefore, linguists are confronted with the challenge of finding common elements around which languages are close enough so as to be comparable. In fact, relevant differences between languages can only be observed insofar as the latter are compared on the basis of a similar concept or structure. If the objects compared differ in nature, then the differences observed will not be relevant. Let us take a practical example. Observing that mice are smaller than elephants is irrelevant to understanding the morphology of mice or elephants, since these are different animals. On the other hand, observing the differences in size between a Chihuahua and a Saint Bernard is relevant for understanding the different morphologies of dogs.

Contrastivists call this point of comparison between languages *tertium comparationis*. Such a point of comparison should be determined in a neutral manner in relation to the functioning of one language or the other, in order not to bias comparisons. For example, comparing the phonological system of French and German using a list of German phonemes as a starting point would provide a biased comparison, since the comparison platform is not neutral but established on the basis of one of the language's categories. It is therefore necessary to choose a point of comparison that can be applied to both languages and, which is, as far as possible, neutral. For example, when trying to compare tense categories between German and English, Gast (2012) selected different time spheres along a time axis, including the *Past Tense*, the *Present Perfect*, the *Present Tense* and the *(will) Future*, independently from both languages. He then drew a line corresponding to the time interval that each tense category covered in each language. Through this comparison, he showed that the English *Past Tense* and the German *Präteritum* seem to cover similar time intervals, whereas the English *Present Perfect* and the German *Perfekt* do not have the same function. Thus, while the English *Present Perfect* only applies to events in the near past, the German *Perfekt* covers a wider range which also includes the distant past, as illustrated in Figure 4.2 (adapted from Gast 2012, Figure 5).

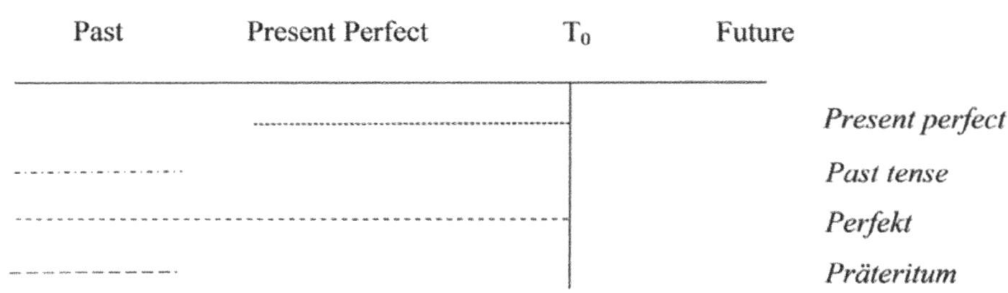

Figure 4.2. *Tertium comparationis for past tenses in English and German*

The choice of the *tertium comparationis* is all the more important since, depending on the chosen point of view, languages may appear to be rather similar or rather different. In his discussion, Krezeszowski (1990) picked the example of squares and rectangles. If these two shapes are compared in terms of their number of sides and angles, they will appear to be identical. However, if they are compared from the point of view of the length of their different sides, they will appear different. The same applies to languages. If French and German tense categories are compared from the point of view of the existence of different tenses for expressing the past, the present and the future, their tense categories will look quite similar. On the contrary, if the comparison concerns the possible uses of the present for designating different temporal references, these two languages will look quite different, since in German the present form of the tense is used for expressing future events, something which in French is expressed in the future form.

The suitable *tertium comparationis* type for carrying out a study depends on the kind of linguistic elements compared (phonemes, syntactic structures, speech acts, etc.). A distinction can be made between the *tertium comparationis* based on linguistic forms and those based on linguistic function (Gast 2012). In terms of the comparison of functions, some focus on the formal correspondence between functions, for example on the existence of certain categories and syntactic functions, whereas others focus on their semantic equivalence, that is, on the similarity of meanings they make it possible to express (Chesterman 1998).

A *tertium comparationis* determined exclusively by formal criteria, however, is not appropriate, not even for comparing structural elements from different languages. As we have seen previously, English and German have two verbal tenses to refer to the past. Thus, from a structural point of view, we could say that these two languages are similar. However, uses between the two languages are quite different. Conversely, a language may lack a certain linguistic form but still express it through other means. For example, in some languages, speakers verbalize the source of information (which they have acquired either directly by their own perception or indirectly by inference or hearsay) by means of a verbal suffix. These languages have what is called an *evidential* verbal system. This is not the case in French, which does not have such suffixes in its verbal morphology. However, French speakers have other means of indicating sources of information in their statements, in particular by adding phrases such as *il paraît que* (it seems that), *j'en conclus que* (I conclude that) or *je vois que* (I see that). So, to infer from the absence of a suffix that the French language does not make it possible to express belief sources would therefore be wrong. That being said, the fact that languages express certain concepts by different means can, in certain cases, give rise to interesting differences, particularly at the age when these elements are acquired by children and the way in which speakers encode this information. The potential impact of such encoding differences on speaker's cognition is known as *linguistic relativism* (see Deutscher (2011) for an argument in favor of the existence of relativism and McWhorter (2016) for a refutation of such).

In many cases, a *tertium comparationis* based on semantic equivalence appears to be preferable to a *tertium comparationis* based on formal criteria.

However, Krezeszowski (1990) warns us against using translation equivalences as an index of semantic equivalence. As we will see in the next section, translations are not always texts entirely representative of a language. In section 4.3, we will provide some in-depth examples of studies that have used a *tertium comparationis* of the semantic type. Let us bear in mind that for certain research questions, notably in the field of pragmatics and discourse, semantic equivalence is not always appropriate. In fact, while two linguistic forms may be semantically equivalent between languages, they may not be used for achieving the same function. For example, in French, it is very frequent to formulate a request indirectly using a question that refers to the interlocutor's capacity, such as in *Peux-tu me passer le sel?* (Can you pass me the salt?). This same strategy is frequently used in other languages such as English and German, but is not universal. For Polish or Russian speakers, a request formulated in this way would not be understood since this typical association does not exist. As a matter of fact, more direct methods for formulating requests are preferred (Ogiermann 2009). Other differences between languages and cultures are discussed by Jaszczolt (2003) in her article on semantic and pragmatic equivalences between languages.

In summary, in addition to being based on corpora with high comparability, contrastive studies should use neutral points of comparison that make it possible to establish comparisons between linguistic phenomena across languages, which are as relevant and adequate as possible. Depending on the research question, the appropriate equivalence levels will be different.

4.3. Translations as a discursive genre

The main question raised by the use of parallel corpora concerns the status of translations and, more specifically, the possibility of using them as language samples. An important amount of research carried out since the 2000s has shown that translations represent a discursive genre in their own right, and that translations do not fully share the same properties as texts written in original language. This discursive genre is also sufficiently stable and different from others so as to be identifiable using machine learning algorithms for automatic text classification (Ozdowska 2009; Ilisei *et al.* 2010).

One of the reasons why translations represent a stylistic genre different from original texts is that these keep a certain imprint of the source language. Even if translators are language professionals, their lexical, grammatical and stylistic choices are still influenced by what they have to translate. For example, Zufferey and Cartoni (2012) observed that the causal connective *since* is used five times less in English to French translations than in original French texts sharing the same register. The reason for this is that this connective is specific to French and has no exact translation equivalent, even in close languages including English (Degand 2004; Pit 2007). This lack of equivalent means that English text translators are much less likely to use it than an author writing in French. There are many other examples of interference created by the source language in translations.

In addition to these influences, which vary from one source language to another, some authors have hypothesized that translations are so similar (to the point of making up a stylistic genre of its own) due to certain effects related to the translation process itself. These effects might reflect translation universals rather than the variable effects pertaining to the languages involved (Baker 1993; Laviosa-Braithwaite 2009). Over the past 20 years, several potential universals have been discussed in the literature. One of these concerns the tendency of translations to be lexically and syntactically simpler than original texts in the target language (Laviosa-Braithwaite 1997). Another universal concerns their tendency to contain a more standardized, less inventive use of the language than original texts (Baker 1993). In the literature, this universal has been linked to the desire of translators to conform to the standards of the target language as much as possible, in order to produce correct texts, something which hinders their creativity in comparison to authors writing in their original language, who can take more liberties. Finally, another universal concerns the tendency of translations to be more explicit than original texts and, more specifically, to contain a greater number of cohesive markers (Blum-Kulka 1986). In the literature, this universal has been explained by the translators' desire to optimize the readability of translated texts by making explicit the type of coherence relations linking discourse segments. In section 4.5, we will present a study that empirically tested the existence of an explicitation universal by means of a parallel corpus.

For all of these reasons, it is important not to use a parallel corpus as if it were a comparable corpus only containing excerpts in the original language. Despite this limitation, parallel corpora represent valuable resources for a number of research questions in contrastive linguistics. As a matter of fact, these are the only data that make it possible to establish equivalences between words or expressions in two languages. In addition, comparing words and their translations makes it possible to better understand the whole semantic field that each word or expression covers in a language. In this case, translations act as a mirror reflecting certain properties of the source language (Noël 2003), which are not always visible in monolingual studies. For example, thanks to this technique, Cartoni *et al.* (2013a) identified six different meanings for the English connective *while*, each corresponding to specific translation equivalents in French. The result of such analysis using the *translation spotting* technique also revealed that numerous occurrences of *while* simultaneously expressed a temporal and a contrastive relation, matching the connective *alors que* in French. In contrast, in dictionaries, the temporal and contrastive meanings of *while* are generally presented separately and appear to be mutually exclusive.

To conclude, in order to limit the bias introduced by the use of translations, it is desirable to use bi-directional corpora as far as possible, as well as to study language equivalences in the two directions of translation. We will see examples of such corpora later in this chapter. We will illustrate the fact that these corpora can help us to work simultaneously on comparable and parallel data, and thus exploit the advantages of each, while limiting their bias.

4.4. Multilingual corpora and contrastive linguistics

In its beginnings in the 1950s, contrastive linguistics emerged as a discipline aiming to compare two or more languages with the aim of improving language teaching methods. Indeed, linguists working on language teaching had long observed that mistakes made by learners were often linked to transfers from their mother tongue. This observation justified the systematic study of differences between languages in order to better understand the risk of making mistakes in different learner populations (see, in particular, Lado (1957)). However, many studies in the field of language learning quickly showed that learner mistakes were by far not always associated with differences between their first and their second languages. On the one hand, in certain cases, gaps between languages should lead to transfer effects, which nonetheless do not take place. Conversely, learners produce numerous mistakes, which cannot be explained through transfer phenomena (see Ortega (2014, Chapter 3)) for a detailed discussion of this).

These new data led to a relative abandonment of contrastive studies for several decades. The situation has changed a great deal since the 1990s, thanks to the arrival of corpus linguistics, which made it possible to empirically compare linguistic systems. The data provided by these contrastive corpus-based studies are not only useful in theoretical linguistics for understanding how languages work, but may be helpful for other applications, notably for the development of tools such as bilingual dictionaries (see section 4.6). In this section, we will present a sample of studies which illustrate the usefulness of corpora for carrying out contrastive linguistic studies.

The first case study that we will discuss concerns the French–English language pair and, more specifically, how the verbs *faire* in French and *make* in English work, both of which can be used in causative constructions such as *faire rire* or *make believe*. On an intuitive level, it may seem that these verbs share a similar meaning and perform equivalent functions in both languages. However, by means of an empirical study of both comparable and parallel data, Gilquin (2008) showed that these two verbs are not equivalent.

Gilquin's study is based on the PLECI bi-directional parallel corpus, which contains newspaper articles and fictional texts in English and French. This corpus can be useful both as a comparable corpus and as a parallel corpus, as illustrated in Figure 4.1. Within this corpus, all the occurrences of the verbs *faire* and *make* were retrieved automatically using a bilingual concordancer (see Chapter 5, section 5.7). Since these two verbs have a host of other non-causative uses, the occurrences had to be sorted manually in order to retain only the causative constructions. The data obtained included 109 occurrences of the verb *make* and 355 occurrences of the verb *faire*. In order to establish the similarities and differences between these two verbs, it was necessary to annotate a set of potentially relevant syntactic and semantic features that could represent a suitable *tertium comparationis*. Gilquin chose to annotate the type of subject of the causal construction (animate vs. inanimate, nominal vs. pronominal), as well as the type of infinitive verb used as a complement of *faire* or *make* (volitional vs. non-volitional, transitive vs. intransitive).

The results revealed some similarities between the two languages. First, the distribution of occurrences between nominal and pronominal subjects was very similar. Second, the two verbs were mainly complemented by verbs describing concrete actions such as *partir* rather than existential verbs like *exist*. Despite these similarities, significant differences were also observed. To begin with, in terms of frequency, the verb *faire* appeared four times more frequently in texts in original French compared to *make* in original English texts, and this was a first indicator that the role of each is not the same in both languages. Furthermore, verbs used with *make* were much more limited than those used with *faire*. The four most frequent verbs in English (*feel*, *look*, *work* and *think*) represented 25% of the occurrences. By contrast, in French, 12 different verbs were needed to reach this same proportion of occurrences. Conversely, some of the uses of the verb *make* seem much more atypical than the verb *faire*. For example, the verb *make* was mainly used in relation to inanimate subjects, which was not the case with *faire*. In sum, although the two verbs have a partly convergent semantic profile, each of them also has frequent uses that are not found in the other language in a similar proportion.

These semantic differences indicate that the verb *make* might not be the best translation choice for *faire*, and the other way around. In order to empirically determine the percentage of correspondences between two words, a mutual correspondence (MC) value can be calculated. This value takes into account the number of translations by the supposed equivalent word compared to the total number of occurrences, in both directions of translation (Altenberg 1999, p. 254). This value is calculated based on the number of occurrences of the two words in translations, which are respectively denoted as A_t and B_t, and then divided by the number of occurrences of these same words in the original texts, denoted as A_s and B_s, and then multiplied by 100 to get a percentage:

$$\frac{(\mathrm{At} + \mathrm{Bt}) \times 100}{\mathrm{As} + \mathrm{Bs}}$$

In the case of the pair made of *faire/make*, the MC value was 15.4%. Such a low value tends to confirm that these two words are not equivalent. In most cases, the causative construction *faire + infinitif* in French is translated by an English verb carrying the notion of causality, also called the synthetic causative. For example, the expression *faire taire* is often translated using the verb *to silence*. In the case of the verb *make*, its most frequent translations are the verb *make* as well as paraphrases, for English expressions that cannot be literally translated into French. For example, the sentence "it was the very intensity of her devotion that had ***made her give***

him a softness of upbringing…" was translated by adding "***Par un excès de tendresse****, Lady O'Connell l'éleva avec une faiblesse…*".

In a nutshell, this study made it possible to show that two words which may seem close, and which are often described as translation equivalents in reference tools such as bilingual dictionaries, are in fact partially different from each other. Furthermore, these differences can only emerge on the basis of a quantitative corpus study, which highlights the differences in frequency and context of use.

The second study we present in this section was devoted to the analysis of the different factors that influence translations in parallel corpora. To do this, Dupont and Zufferey (2017) compared the way in which concessive connectives are often treated as translation equivalents in bilingual dictionaries, namely: *however*, *yet*, *nevertheless* and *nonetheless* in English and respectively *pourtant*, *toutefois*, *néanmoins* and *cependant* in French. The authors specifically studied the role of three factors in the observed equivalences: the translation direction (French–English or English–French), the stylistic genre (journalistic texts or parliamentary debates) and the translators' degree of expertise (non-professional volunteers, journalists or qualified translators). For this study, the occurrences of the eight above-mentioned connectives were drawn from three parallel corpora (Europarl for the parliamentary debate genre, a corpus of newspaper articles and the TED corpus of online conferences; see Chapter 5 for a description of these corpora). These occurrences were then manually disambiguated in order to remove occurrences which had not been used as a concessive connective, for example when the connective *yet* was used to indicate a temporal relation.

The results showed that in original texts, the frequency of connectives often vary depending on language register, particularly in English, where the four connectives vary significantly. In French, only the connective *pourtant* varied significantly between journalistic texts and parliamentary debates. An analysis of translations also showed differences between the two genres. For French connectives, the typical translations in the journalistic genre were either the generic connective *but*, or there was an outright absence of a connective in the translation. In the parliamentary debate genre, more specific connectives were used: the connective *however* was a frequent translation for the four French connectives, not to mention *yet* as the translation of *pourtant* and *nevertheless* for *néanmoins*. Such a tendency to omit connectives in the journalistic genre can also be found in English. This observation can no doubt be explained by the concern for efficacy in this genre, which tends to limit the amount of words used. The other translations were more variable than in the French–English direction.

We can see that the direction of translation is an important factor to take into account when establishing equivalences between languages. Differences between stylistic genres were also visible in the MC values between connectives. Indeed, these values were very low in the journalistic genre, oscillating between 14% and 27%, against 33% and 57% in the parliamentary debate genre, which reflected the above-mentioned more specific translation choices.

The last variation factor analyzed in this study referred to the translator's level of expertise. On the one hand, European Union translators are qualified professionals. On the other hand, the translations provided for TED conferences are carried out by volunteers. Finally, journalistic text translations are generally carried out by journalists, who are language professionals but not translation professionals. For the English–French pair (remember that the TED corpus is unidirectional), these variations enabled the authors to compare the impact of this variable on the translations under scrutiny. The comparison revealed that translation choices were systematically less varied in the TED corpus than in other corpora. The number of zero translations was also significantly lower. This trend reflected the fact that amateur translators are more likely than others to use the source text as a guide and to avoid structural changes as much as possible (see also Lefer and Grabar (2015) for a similar conclusion). In other words, their translations are often more literal than those of professional translators.

In summary, this study showed that the type of equivalences observed between languages can be variable across discourse genres. However, contrary to what happens in monolingual studies, contrastive studies are often performed on data from a single genre – due to the scarcity of multilingual corpora – which does not always make it possible to compare different genres. This study also showed that equivalences between languages should be considered separately for the two translation directions. Finally, the degree of expertise of translators also plays a role in their translation choices, and this factor should therefore be taken into account in the study of parallel corpora.

4.5. Parallel corpora and translation studies

Translation studies is the scientific study of the processes at work in translation, as well as the factors that influence their realization. While translation is a practical and applied discipline, translation studies (translatology or traductology) is a theoretical science. As in the case of contrastive studies, translation studies has benefited from the availability of multilingual corpora, as well as theoretical and methodological advances in corpus linguistics. As we will see in this section, the use of large multilingual corpora makes it possible to carry out quantitative studies on different language pairs simultaneously and, therefore, go beyond the isolated observations that can be made on the basis of individual practice. Later, we will see that the use of the empirical methodology ingrained in corpus analysis can also work as a guide for the translator when it comes to making certain translation choices.

The first study that we present in this section looked into the existence of translation universals. As discussed previously, translations differ in several ways from original texts produced in one language. Translation studies specialists have suggested that a portion of these specificities can stem from the existence of translation universals, that is, from phenomena specifically pertaining to the translation process. One of these universals concerns the supposed propensity of translations to be more explicit (explicitation phenomenon), in terms of cohesion markers, than original texts. This hypothesis has been partly confirmed through corpus studies, performed on a single language pair and limited to one translation direction. Due to these limitations, these studies cannot be generalized to all translations.

In order to overcome this limitation, Zufferey and Cartoni (2014) used the multilingual corpus of parliamentary debates, Europarl, in order to determine whether explicitation phenomena were evenly observable when different variation parameters such as the source and target languages were tested, while keeping the factors of stylistic genre and translation quality constant across language pairs. The main advantage of using the Europarl corpus to carry out this study is that all languages are alternately source and target, and the texts contained in each portion of the corpus deal with very similar subjects, and they were produced under highly similar conditions (parliamentary debates), which guarantees their comparability.

The explicitation hypothesis relates to the number of cohesion markers present in translations, which is assumed to be higher than in original texts. Among these, Zufferey and Cartoni chose to focus on the category of causal connectives. Indeed, their use is very frequent, and often optional. In other words, they can be omitted without creating comprehension difficulties (Murray 1997), something which makes them perfect candidates for testing explicitation phenomena. If translators tend to make optional cohesion markers explicit then the number of causal connectives should significantly increase in translations, when compared to original texts. The methodological challenge of this study consisted of identifying those cases where a causal connective had been added in a translation. To achieve this, the authors looked for occurrences of the four French causal connectives (*parce que*, *car*, *puisque* and *étant donné que*) in the corpus section containing translated French, and checked whether an equivalent connective was also present in the source text, thus carrying out a form of reverse *translation spotting*. If no causal connective was present in the source text, then this would be an example of explicitation. This technique made it possible to count the number of connectives added to French translations from four different source languages: English, German, Spanish and Italian. The results showed that there were many cases of explicitation in translations (connectives had been added despite the lack of any source language indicator in about 7% of the cases), but this rate did not vary significantly depending on the source language. The authors then changed the target language, looking for the three causal connectives, *because*, *since* and *given that* added in English texts translated from French, German, Italian and Spanish. Once again, they observed the recurrent presence of explicitation phenomena but this rate did not vary, regardless of the target language. These results provided a first hint of evidence that explicitation was indeed a regular phenomenon in translations, regardless of the language pair involved.

Furthermore, the authors were able to observe that the explicitation rate varied significantly depending on the causal connective in question. On the one hand, some connectives like *parce que* in French and *because* in English gave rise to very few explicitation cases. On the other hand, causal connectives like *puisque* in French and *given that* in English gave rise to many explicitation cases. The authors attributed this gap to the different semantic profiles of connectives. Those that give rise to explicitation are typically used for introducing a cause presented as already known or easily inferred by the interlocutor, unlike the other connectives which are used for announcing a new cause for the interlocutor (see Zufferey and Cartoni (2012) for a contrastive study of causal connectives in French and English). This observation reinforces the idea that explicitation reflects the desire of translators to improve text readability, by explicitly showing readers that a piece of information is considered by the author to be already known by the audience or easily accessible.

The second study that we will discuss does not deal with the analysis of translations themselves but with the stylistic analysis of the source text, namely the search for recurring *patterns* and monitoring how these *patterns* are translated. Čermáková (2015) studied the recurring stylistic elements in John Irving's novel *A Widow for One Year* and the way in which these were translated into Finnish and Czech. As we saw in Chapter 3 (section 3.6), corpus linguistics provides analytical tools that are very useful for the study of literary texts. In particular, they help to identify the keywords of a text and to analyze them in context. In the case of literary translation, the author argued that a preliminary stylistic analysis of the translatable material made it possible to identify certain recurring *patterns* that were not easily identifiable through qualitative research, and justified the need to treat them in a systematic manner. Using a concordancer, she analyzed the repeated sequences of words in Irving's novel and found eight-word sequences that were repeated at least three times; 27 sets were identified. She also generated a list of keywords in the novel, using the *British National Corpus* as a reference corpus. A comparison between the word sequences and the keywords revealed that most of the recurring word sequences contained or referred to a keyword (see Chapter 5, section 5.7 on methods for generating a list of keywords).

By analyzing the recurring sequences and the keywords they contained, the author was able to show that these repetitions played a particularly important stylistic role in the novel (which also contains many more repeated sequences than other works by the author), and that these repetitions should be maintained in the translation in order to preserve the spirit of the text. In fact, these sequences made reference in part to the titles of other literary works, and helped to grasp certain intertextuality elements. However, an analysis of the translations of these 27 recurring sequences, both by the Finnish translator and by the Czech translator, showed that they were mostly neutralized by stylistic choices avoiding repetitions. The tendency of translators to avoid repetition is also one of the recurring trends identified in translations, and some translation theorists point to various techniques for achieving that result (Ben-Ari 1998). However, as in the case of Irving's novel, the presence of repetitions may be an integral part of the work's style and erasing them would certainly involve a form of stylistic loss.

In this way, we can see how corpus linguistics can provide translators with tools that may help them adapt their translation choices on the basis of a better identification of the recurrent linguistic properties at work in a text.

4.6. Parallel corpora and bilingual dictionaries

We have already discussed the importance of corpora for monolingual lexicography in Chapter 3 (section 3.5). In this section, we will refer more specifically to the role of parallel corpora in the creation of bilingual dictionaries. Bilingual dictionaries are essential for foreign-language learners but they are also controversial among language professionals, especially translators. The latter, in particular, criticize bilingual dictionaries for the limited list of equivalences that they provide and the lack of context, which often prevents users from making an appropriate distinction between the different meanings of a word or expression. Finally, as monolingual dictionaries, these dictionaries do not provide any indication regarding the frequency of the different meanings, apart from the order in which they are listed.

To some extent, equivalences between languages obtained through the use of parallel corpora respond to such criticisms. Corpora provide access to a broad context and offer a greater variety of equivalences than dictionaries. What is more, these can be easily classified by frequency, and differentiated according to the textual genres under consideration. In addition, computerized word alignment techniques make it possible to automatically produce bilingual dictionaries (see, for example, McEwan *et al.* (2002)). These same techniques also inspired online dictionaries such as *Linguee*[1] bilingual dictionaries, a resource which is based entirely on parallel corpora drawn from the Internet. The huge advantage of these resources is the diversity of translations they offer and the broad context that accompanies each of them. However, as translations are automatically identified, their accuracy is not guaranteed but requires a critical evaluation on the part of users.

---

1 Available at: https://www.linguee.com/.

In order to illustrate the importance of corpus data for providing more suitable translation equivalents than those of bilingual dictionaries, in this section, we will discuss a study concerning partially equivalent word pairs in French and in English. Cummins and Desjardins (2002) studied the different meanings of the words *population* in French and *population* in English, as well as fixed expressions such as *plus au moins* in French and *more or less* in English. The authors found that bilingual French–English dictionaries listed these words and expressions so as to convey the idea that these were completely equivalent. Then, resorting to the main monolingual dictionaries, they set up a list of their possible meanings in French and in English. When comparing the two lists, the authors realized that some of the meanings could not be found in the other language. For example, only the French use the term *population* in an emotional sense and in political contexts, and the expression *plus au moins* with a euphemistic sense.

At a second stage of the study, the authors looked for occurrences of these words in French–English comparable corpora. They chose 100 occurrences in each language and annotated them with the different meanings listed in monolingual dictionaries. They confirmed that some of the meanings frequently found in the corpus could not be adequately translated by their "equivalent". For example, 75% of the occurrences of *plus au moins* in the corpus should have been translated using expressions such as *pretty much* or *somewhat* in English, rather than using the expression *more or less*. The authors concluded that bilingual dictionaries do not provide enough information for helping users access the correct translation equivalents.

Many other studies have compared the translation equivalents provided by bilingual dictionaries with equivalents observed in parallel corpora. These studies invariably highlight a discrepancy between the translation equivalents found in dictionaries and in corpus data. In most cases, the equivalents provided by dictionaries are much more limited than the equivalents found empirically, or vice versa, dictionaries sometimes list equivalents that are completely absent from corpus data. We will work on two examples by way of illustration. Degand (2004) studied the causal connectives *puisque* in French and *aangezian* in Dutch, which are treated as equivalent in bilingual dictionaries. However, *puisque* was only translated as *aangezian* in 42% of the occurrences in parallel corpora. An even more striking result, *aangezian* was only translated as *puisque* in 8% of the cases. In another contrastive study on the French and English causal connectives,

Zufferey and Cartoni (2012) found that *puisque* was translated as *since* in only 43% of cases and that *since* was translated as *puisque* in only 23% of cases, whereas these two connectives are usually presented as equivalent in bilingual dictionaries. These examples illustrate the need to integrate corpus-based data in bilingual dictionaries in the future, in order to provide users with a more empirically based view of equivalences between languages.

4.7. Conclusion

In this chapter, we have discussed the different uses of comparable and parallel multilingual corpora. We have seen that their advantages and disadvantages are often complementary, and that it is useful to combine these two types of resources in contrastive linguistics. The study of translation often relies on parallel corpora, but can also make use of comparable corpora of texts translated into different languages, without considering the source language. In the field of translation studies, one of the major aims of such studies is to analyze the features of the translated language, with the purpose of looking for translation universals. We have also shown that corpus analysis methods can be useful for uncovering recurring patterns in a source text and to better adapt the strategies used for its translation. Finally, we argued that parallel corpora have become indispensable resources for the creation of bilingual dictionaries, since they provide rich lists of translation equivalents accompanied by their contexts of use, as well as information concerning their frequency in various genres.

4.8. Revision questions and answer key

4.8.1. *Questions*

1) What type of multilingual corpus (comparable or parallel) seems most suitable for studying the two research questions stated below?

a) What are the similarities and differences between the causal connectives *porque* in Spanish, *parce que* in French and *perché* in Italian?

b) How are the European elections reported in the press in Germany, France and the United Kingdom?

2) What would be a good *tertium comparationis* for the following two research subjects?

a) Comparison of the German and the French consonant systems.

b) Speech acts of thanking in French and Chinese.

3) Why can we say that translations are a full-fledged text genre?

4) What are the parameters to take into account in order to carry out a contrastive study on the use of the indefinite pronouns *on* in French and *man* in German?

5) How could we test the supposed translation universal according to which translations are simpler than original texts by means of a parallel corpora study?

6) What types of equivalences are most likely to be insufficiently dealt with in bilingual dictionaries?

4.8.2. *Answer key*

1) a) In order to study the **similarities and differences between the causal connectives *porque* in Spanish, *parce que* in French and *perché* in Italian**, the use of a parallel corpus offers great advantages. Indeed, such a corpus makes it possible to establish the degree of mutual correspondences between these connectives, by counting the number of times that they can be translated by each other. Nonetheless, the use of this method also involves the risk of having a distorted vision of the functioning of these connectives, due to the translation prism. This study should therefore be supplemented by a semantic and pragmatic analysis on how these connectives work in the original language, by means of comparable corpora. For instance, the use of these connectives could be compared only in the source language section of the parallel corpus.

b) Conversely, to study **the way in which European elections are reported in the press in Germany, France and England**, the use of comparable corpora seems the most judicious choice. Indeed, for this study, it is important to have access to texts that were originally produced in each language, in such a way that they reflect both the linguistic structures of each language and bring out potentially different discourses regarding the same event. A parallel corpus, containing translations, would not be able to meet these two objectives.

2) a) The **comparison of the consonant system in German and French** can be done on the basis of formal rather than functional equivalences. In particular, consonants can be compared on the basis of their articulatory features.

b) In order to compare **speech acts of thanking in French and Chinese**, a *tertium comparationis* based on pragmatic equivalence is necessary. It is not only the words or expressions that should be compared, but also their illocutionary force, that is, the communicative intention of the speaker.

3) Several reasons have been given in the literature for explaining the linguistic specificities of translations. The first type of explanation concerns the influence of the source language, which inevitably leaves traces in translations. Even if translators are language professionals, they are inevitably influenced by the words and linguistic structures they have to translate, which leads them to make different lexical and syntactic choices than those of a speaker writing in their mother tongue. The second category for explaining translation specificities is of a general nature and is based on the supposed existence of translation universals (linguistic phenomena resulting from the very process of translation), regardless of the source and target languages involved. These universals include simplification, explicitation and standardization. All these processes reflect the pedagogical role of translators, who (unconsciously) try to improve the readability of texts.

4) First of all, this study should be carried out by means of a parallel corpus, in order to determine to what extent these two pronouns are translation equivalents or not. More specifically, a bi-directional parallel corpus should be used, since the equivalences are often variable depending on the direction of translation. This analysis of translations should be supplemented by a study on comparable corpora, made up of the two original language sections from the parallel corpus. For this analysis, the important point would be to establish which comparison factors would best highlight their common points and their differences. In this case, the possible factors could be the tense and aspect of the verb following the pronoun, etc. Finally, this study should, wherever possible, include two different discourse genres, in order to measure the extent of the variations between them.

5) The simplification universal implies that translations should be simpler linguistically than the original texts of the same discursive genre. Various lexical and syntactic factors, easily measurable, could contribute to this simplicity. For example, lexical simplicity implies that the number of different words should be smaller than in an original text. This can be measured thanks to the type/token ratio (see Chapter 8). Syntactic simplicity is measured, for example, by the average length of sentences. The number of words used per sentence can also be calculated, even on a corpus that has not been subjected to syntactic annotations.

6) The most problematic equivalence cases for bilingual dictionaries are partial equivalences, just as those we discussed in this chapter for expressions such as *plus au moins* and *more or less*. In these cases, the formal proximity and the identification of certain cases in which these expressions are equivalent may suggest that these words are completely equivalent, when actually they are not. Conversely, false cognates, where meanings are completely different between languages despite a formal resemblance, are easier to identify, since their meaning clearly appears to be different.

4.9. Further reading

Kenning (2010) provides a concise presentation of the similarities and differences between comparable and parallel corpora. The book edited by Sharoff *et al.* (2016) contains many chapters dedicated to the construction, evaluation and use of comparable corpora. Johansson (2007) is an essential reference on the use of multilingual corpora in contrastive linguistics. The question of translation universals is discussed in detail in the work by Mauranen and Kujamäki (2004). The different uses of multilingual corpora for contrastive linguistics and the study of translations are discussed in an accessible way by Mikhailov and Cooper (2016).

5

How to Find and Analyze Corpora in French

This chapter has two aims. Firstly, we will introduce the main existing corpora in French. These corpora can be divided into four categories:

– written corpora;

– spoken corpora;

– corpora devoted to specific demographics, such as children or learners;

– multilingual corpora where one of the languages included is French.

Secondly, we will present a set of concordancers, which are corpus analysis tools, and discuss their main functionalities. Links to websites providing online access to corpora, as well as corpus consultation tools, are listed at the end of the chapter.

5.1. Corpora formats and their availability

Thanks to the Internet, in recent decades sharing corpus data has become far simpler. For example, it is very common for research teams to offer the corpus compiled during their research projects as a tool available to the general public, once the project is finished. The rights of access and use of this data may vary depending on the content and project in question. In some cases, corpora available to the public can be downloaded directly from a website. In other cases they are not downloadable, but instead are only available online, via a dedicated search interface. We will discuss the advantages and disadvantages of these different formats in this section. It should be noted that use for commercial purposes may be subject to specific additional conditions.

The main advantage of downloadable corpora is their great flexibility for carrying out word or structure searches using a concordancer (see section 5.6). These corpora can also be annotated manually or automatically. An important element to take into account when downloading a corpus from the Internet is its encoding format (this not to be confused with the compression format, when applicable, that needs to be handled by means of specific archiving tools). When several formats are available, it is important to choose a format which is compatible with the research tools that will be used for the study. For example, the AntConc concordancer that we discuss can only process text format files (or files with XML or HTML tags which can also be treated as text files). Another element to take into consideration before deciding to download an entire corpus is its size. Indeed, some current corpora like the *Google Books* corpus (Michel *et al.* 2001) are extremely large, making them nearly impossible to use from a private computer, even if they are available for download. In these cases, it is preferable to use the online consultation tools or to only download portions of the corpus, as needed.

Many corpora can only be viewed online using a dedicated research interface. The advantage of this format is its great simplicity of use. In fact, most interfaces offer user-friendly methods specifying the choice criteria, such as gender, type of speaker, time period, etc., as well as fields for typing in the element(s) to be looked for in a full text search, sometimes enabling the use of search patterns (called regular expressions, see section 5.6). The major drawback of these interfaces is that they do not authorize any type of search. Some are limited to a continuous character string, something which prevents the search for compound words, like *chemin de fer* (railway) in French, which includes three separate strings of characters. If the search patterns are not usable, this further complicates the search. Let us take a look at an example: it is possible to look for all the occurrences of a regular verb like *aimer* using a single query looking for the root *aim*, followed by a wildcard replacing an unspecified number of characters, for example *aim**. If this type of search is not enabled by the interface, all the verbal forms must be looked up one by one with their exact forms.

Another limitation to the use of online corpora available for consultation is that they only offer limited access to their metadata (see Chapter 6). Thus, important information for certain research questions is regularly missing. For example, the online interface on the *OFROM* corpus (see section 5.3) only offers the possibility of looking for productions by women or men, but for the moment does not indicate the total number of participants of each genre, or the number of words produced by each of them. This type of information is nonetheless crucial for the quantitative comparison of linguistic productions between the two genders. In the same way, the consultation interface for the *CLAPI* corpus (see section 5.3 for a description) does not currently provide information about the total number of words included in the portion of the corpus that is available for online consultation. In many cases, this piece of information can be obtained by contacting the corpus creators. However, when this type of information remains unavailable, these gaps lead to serious limitations in data analysis (see Chapter 8). The problem of sources is even more acute in the case of databases grouping different types of corpora, such as the Lextutor database, which contains both spoken and written data, retrieved from different genres, but unevenly distributed. These databases are useful for quickly finding concordance examples but cannot be seriously considered as representative corpora (see Chapter 6).

In addition to the two above-mentioned distribution formats, some corpora that are available online require prior user registration, as well as explicitly stating the research purpose for which the data will be used. For example, this is the case of the Belgian Valibel database of spoken French (see section 5.3) or the SMS corpus in Switzerland's national languages, collected by the universities of Zurich and Neuchâtel (see section 5.5).

Other corpora are still not distributed for free but can be obtained by paying a varying fee, depending on whether the intended use is for research or for commercial purposes. After purchase, *Le Monde* newspaper corpus (see section 5.2) can be downloaded via corpora distribution sites such as the European Language Resources Association or ELRA, the Linguistic Data Consortium or LDC, or distributed in a CD-Rom format, such as the French SMS corpus collected in Belgium. Other corpora such as the new version of the *Frantext* literary text corpus (see section 5.2) are accessible via an annual renewable subscription. Many corpora are also available via the Sketch Engine online platform (see Chapter 6), which is free to access for many institutions in European countries. In addition, institutions often finance the purchase of corpora for their members, so it is advisable to check with one's institution before engaging in any individual purchases.

Finally, we can mention that it is possible to build new corpora from websites that distribute royalty-free data, for example, literary texts now available in the public domain, government data such as parliamentary debates, or texts from participative sites like Wikipedia. In all cases, it is necessary to study the rights of use indicated by the respective sites before starting to compile the corpus. We will discuss this issue in more detail in Chapter 6, which is devoted to presenting the basic principles of corpus creation.

To conclude, it is important to emphasize that, regardless of the format in which a corpus can be accessed, reusing corpus data amounts to benefiting from the often long and costly work carried out by other research teams. That is why, when existing corpora are used, their source must be explicitly mentioned. More specifically, when researchers reuse a corpus created by other teams, they must mention in their publication the Internet link where the data were downloaded or retrieved from. Very often, the authors of a corpus provide a bibliographic reference where they describe their corpus or the name of the team who compiled it. These references must be quoted in any paper making use of the data.

5.2. Reference corpora

Unlike many European languages, French still does not have a reference corpus, a representative sample of the French language in general, similar to the *British National Corpus* that exists for British English, one of the pioneers in the genre. For the time being, it is not possible for linguists to observe how the French language works through the study of a single corpus. However, a multi-genre French corpus is currently in development and should offer an open access phase to the general public in the coming years (Siepmann *et al.* 2016).

For the time being, the closest to a reference corpus for French is the corpus of contemporary French created within the framework of the Orféo project (Benzitoun *et al.* 2016) that brings together existing written and spoken corpora pooled by various research teams. Available online, this corpus currently includes 15 spoken corpora that amount to approximately 4 million words and six written corpora, comprising 6 million words. The online interface makes it possible to launch requests throughout the entire *Orféo* corpus, or to create a sub-corpus by choosing the mode (spoken or written), genre (literature, press, etc.), and for the spoken corpus, the interaction mode and the number of speakers. Another advantage of this resource is that the corpus it includes has been enriched with morphosyntactic annotations (see Chapter 7), making it possible to refine the search criteria.

The Sketch Engine corpus management system (Kilgarriff *et al.* 2014) brings together corpora in many different languages, including French. These data include open access corpora on the Internet, such as the CHILDES database (see section 5.5) and the French section of the Europarl corpus (see section 5.6), as well as several large corpora assembled using web crawling techniques. For French, this corpus is called French Web 2012 (which is part of the *TenTen12* series of corpora in different languages; see section 5.6), containing almost 10 billion words. In addition to the existing resources, Sketch Engine offers the possibility of creating and managing new corpora either from automatic web crawling or by inserting files (see Chapter 6), which further increases the interest of the resource. Many universities offer access to the Sketch Engine database, and access for individual researchers is available via an annual subscription.

In addition to the big generic corpora, for certain types of research, using corpora belonging to a specific type of genre may prove to be a wise choice. The results obtained from different specific corpora can also be combined in order to improve the generalization of results. We will describe these corpora in the following sections.

5.3. Written French corpora

In the field of journalism, the most exhaustive resource is undoubtedly the *Le Monde* corpus, which contains the newspaper's archives for the period 1987–2012, representing a total of nearly 1,200,000 articles, corresponding to almost 20 million words per year. The *Le Monde* corpus is a valuable tool not only for exploring the French journalistic style but also for studying recent developments in the language, thanks to its data spanning 25 years. Unfortunately, this corpus is not available for free and must be purchased via the ELRA platform. On the other hand, the newspaper's articles for the year 1998 are available for free online consultation via the Lextutor platform.

Another newspaper is also a good reference for the French journalistic style. The *Corpus Journalistique issu de l'Est Républicain* includes articles from this regional newspaper for the periods 1999–2003 and 2006–2011. The corpus can be downloaded for free from the Ortolang corpus distribution platform. The *French Discourse Tree Bank* (Danlos *et al.* 2012) is a corpus of texts drawn from the same journal for 1992, which includes syntactic and discursive annotations.

In the more specialized journalistic genre, the *Sciences Humaines* corpus produced by ATILF (*Analyse et Traitement Informatique de la Langue Française* [Computer Processing and Analysis of the French Language]) in Nancy includes 125 linguistic articles from the *Sciences Humaines* journal. Despite its modest size, this corpus makes it possible to study the specificities of the journalistic style when applied to a particular field. The corpus is now available via the Ortolang platform, where it can be downloaded for free.

As for the literary genre, the *Frantext* corpus brings together many literary texts ranging from ancient to modern French, in a corpus which totals more than 250 million words. Since 2018, a new version has made it possible to consult the corpus by means of an improved interface, facilitating the search for regular expressions. The corpus has been lemmatized and tagged into grammatical categories, which also helps in refining the search criteria. This version of the corpus is available online but requires a paid subscription. A portion of the corpus, including works from the 18th to the 20th Century, can be downloaded for free from the Ortolang website. The site's interface makes it possible to choose works based on different criteria, such as the time period or the author.

The *Base du français médiéval* (Guillot-Barbance *et al.* 2017) offers access to different diachronic corpora. The main corpus, BFM 2016, includes 153 texts, corresponding to more than 4 million words. This database also provides access to the *Corpus représentatif des premiers texts français* or CORPTEF, which brings together texts from the 9th to the 12th Centuries, and to the *Passage du latin au français* corpus or PALAFRALAT, which aims to document the linguistic transitions between Latin and French. These data are available free of charge and can be viewed through an online interface. Most of the texts can also be downloaded in PDF format.

The *Google Books* corpus (Michel *et al.* 2011) contains more than 5 million digitized books in different languages, including 45 billion words in French. While the corpus itself remains entirely private, an online consultation interface makes it possible to find the number of occurrences of individual words or word sequences (Ngram viewer), also specifying a desired time interval. The corpus has also undergone a part-of-speech tagging process which makes it possible to only look up words within a certain grammatical category. Other tags can be used for looking up a word at the beginning or at the end of a sentence only, or to look up words regardless of their morphological inflections (lemmatized form). Frequency information for all sequences (Ngrams) of up to five words can also be downloaded, which can be useful for certain searches which exceed the possibilities provided by the interface (see Chapter 2, section 2.4 for an example on how to use this corpus). The main limitation of this interface is that it offers information concerning word frequencies or expressions across time but does not provide the context for each occurrence, as a concordancer would.

In the field of new media, the CoMeRe database includes communication corpora mediated by networks, such as SMS/text messages, *tweets*, blogs, etc. These data are accessible via the Orféo platform. Also in the field of new media, the Belgian *sms4science* corpus (Fairon *et al.* 2006) includes a collection of over 75,000 text messages collected in the 2000s, produced by 3,200 people. The messages come from different French-speaking regions: Belgium, French-speaking Switzerland, Quebec and Reunion. The number of words differs significantly from region to region. According to Cougnon (2015), the corpus includes nearly 700,000 words for Belgium, approximately 233,000 for Reunion, 61,000 for Quebec and 94,000 for Switzerland. This corpus can be bought as a CD-Rom and is accompanied by an introductory book. A French component was collected in the region of Montpellier (Panckhurst *et al.* 2013). This corpus, called *88milSMS*, is available on request from the authors. Finally, the *Presidential2017* corpus, produced by the Agora laboratory at the University of Cergy-Pontoise, contains an archive of tweets produced during the 2017 French presidential campaign. This corpus contains a total of almost 45,000 tweets. It can be downloaded for free from the Ortolang website.

Finally, the *Corpus Français de l'université de Leipzig*, which is not actually a corpus *stricto sensu* as it contains a set of isolated sentences rather than whole texts, brings together different sources such as newspapers and web pages, as well as entries from the participative encyclopedia, Wikipedia. The data collection mode makes this corpus unsuitable for many types of research but provides a very useful interface for lexical searches, offering the possibility of looking for simple or compound words and having access to all the occurrences within the context, with an indication of the source for each occurrence. The interface also automatically generates a list of the most frequent co-occurrences for each word, as well as the most frequent words to the left and to the right of the word in the search. For each request, the interface returns an indication of the frequency rank of the word looked up in the corpus. This piece of information makes it possible to estimate the potential difficulty of a word, for example, in the context of language teaching or for preparing experimental material, by controlling the frequency of the words used in the experimental materials.

5.4. Spoken French corpora

Numerous spoken corpora have emerged since the 2010s. Here, we limit our presentation to resources of a general nature, which are at least partly publicly available. However, many other more specific resources can also be downloaded for free from the Ortolang platform.

The corpus of spoken languages in interactions or CLAPI (Groupe ICOR 2008) brings together the transcripts of approximately 40 social interactions filmed in a natural context. These interactions correspond to different social situations, both professional and private, such as business transactions, teaching sessions, a drink with friends, etc. The corpus site provides free access to 46 hours of interaction, corresponding to the transcriptions of 140 dialogues. The search interface also allows you to choose only certain types of interaction, as well as the number of participants and the presence or absence of non-native French speakers. Requests may only concern strings or discursive phenomena such as overlaps, pauses, etc., and the results can be checked later, not only through transcripts, but also by accessing the audio-visual recording.

The *Corpus oral de français de Suisse romande* or OFROM corpus was collected at the University of Neuchâtel (Avanzi *et al.* 2017) and includes transcriptions of speaker recordings from all the cantons in French-speaking Switzerland, which were aligned with the audio file. In total, the corpus includes more than a million words, produced by more than 340 different speakers. The corpus can be consulted via an online interface, which makes it possible to look up words or word sequences, specifying their grammatical category, as well as left/right contextual information, if needed. The interface also makes it possible to specify sociolinguistic criteria such as gender, socio-educational level and age of the speakers.

The *Corpus de français parlé parisien dans les années 2000* or CFPP2000 (Branca-Rosoff *et al.* 2012) was collected at Sorbonne Nouvelle Paris 3 University. It puts together a set of long and non-directive interviews about the way people perceive their neighborhood. This corpus comprises approximately 54 hours of recordings corresponding to nearly 700,000 words whose transcriptions have been aligned with sound. The corpus is accessible free of charge via an online interface, making it possible to look up words in transcripts, as well as to specify certain sociolinguistic criteria such as gender, age and speaker's mother tongue.

The *Corpus de français parlé au Québec* or CFPQ was collected at the University of Sherbrooke (Dostie 2012) and aims to reflect the French spoken in Quebec in the 2000s. It is a multimodal corpus offering a broad range of information, spanning from the linguistic, prosodic, vocal (laughter, sighs) to the gestural (applause, imitations). This corpus consists of nearly 700,000 words and is freely accessible via an online interface. This interface makes it possible to look up words or word combinations, as well as to specify sociolinguistic criteria, such as gender, age or educational level.

The Belgian French Valibel corpus (Dister *et al.* 2009) was compiled at UCLouvain between 1987 and 1995. It includes 22 different corpora with more than 370 hours of recordings, representing more than 500 different speakers, and nearly 4 million words. This corpus is constantly evolving and new data dating from the 2000s are currently being added. The corpus is available online via the *Moka* interface. However, access is limited to users who require it via an online form.

The *Traitement des corpus oraux en français* project or TCOF from the ATILF laboratory brings together corpora collected between the 1980s and 1990s, and later enriched in the 2000s. The portion of the corpus available to the public not only includes interactions between adults and children, but also interactions between adults only. It contains 124 transcripts of dialogues, ranging from 5 to 45 minutes aligned with the sound, representing a total duration of 124 hours. The CID corpus (Bertrand *et al.* 2008) is also an interaction corpus incorporating phonetic and syntactic annotations. These corpora can be downloaded from the Ortolang platform.

Finally, the Backbone Project contains many videos of interviews with young speakers of different languages, including French. This corpus was designed to document less commonly taught languages or regional varieties of widespread languages. In the case of French, the interviews include young people from the Guadeloupe and Montpellier regions, in particular. This corpus also has an educational purpose in the area of language teaching. This is why it has incorporated grammar, lexicon and language register annotations, which can be looked up through the online interface.

5.5. Children and learner corpora

Many language acquisition corpora can be found in the Child Data Exchange System or CHILDES database (MacWhinney and Snow 1985). All CHILDES corpora have been transcribed in a unified format called CHAT. They are downloadable and searchable using multiplatform software (CLAN) which can be downloaded for free on the same site. Alternatively, these corpora are accessible via an online interface. The CHILDES corpora contain data in many different languages, although a significant portion of the corpora is in English.

A specific section of the CHILDES database is dedicated to French corpora. In 2019, this section amounted to a total of 16 corpora. The majority of them (12 out of 16) are longitudinal corpora, comprising between one and six children. Half of the corpora focus on very early childhood, with speakers between 1 and 3 years of age, a period during which many elements of the spoken language are acquired. The other corpora include children up to the age of 7, and only one of them (*VionColas* corpus) includes children up to 11 years old. Six corpora are only available as written transcriptions, five others also have access to sound, and five of them include a video recording.

The CHILDES database also offers a section on bilingualism, including five corpora for which one of the languages is French, the other language is Portuguese (*Almeida* corpus), Dutch (*Amsterdam* corpus), Russian (*Bailleul* corpus) and English (GNP and *Watkins* corpora). These corpora contain recordings from one to seven children, ranging from 1 to 7 years old. Most of them are longitudinal corpora.

A section of the CHILDES database is dedicated to children with atypical language development. Some of the corpora in this section include data in French. The *FoudonReboul* corpus is a longitudinal corpus of eight children with autism spectrum disorder (ASD), recorded between the ages of 4 and 9 years. The *Nadig* corpus also includes 28 French-speaking children diagnosed with ASD, aged between 3 and 7 years. *Le Normand* corpus includes seven children diagnosed with epilepsy and specific language impairment (SLI), recorded between the ages of 4 and 5 years.

Apart from the CHILDES database, some recent corpora have aimed to study the development of written language in older children. This is the case of the *EMA écrits scolaires* corpus (Boré and Elalouf 2017), a longitudinal corpus bringing together the text productions of primary and middle school children. A portion of the corpus contains narrative texts produced by CP and CE1 students (6–7 years old), while the other section is made up of a series of argumentative texts produced by CE2 and CM1 students (8–9 years old). The corpus includes images of handwritten texts, raw transcriptions in text format and an annotated transcription, also in text format. This corpus can be downloaded from the Ortolang platform.

In the field of written French language acquisition, the *Littéracie avancée* corpus produced by the *Laboratoire linguistique et didactique des langues étrangères et maternelles* (LIDILEM) of Grenoble Alpes University is made up of writings by undergraduate and master's degree students, covering the entire span of study. It contains academic writings such as dissertations, book synopses and reports, as well as motivational letters. The corpus is made up of 11 sub-sections containing at least 10 texts each, produced under similar conditions, namely by students of the same level. This corpus can be downloaded from the Ortolang platform.

In the area of French as a foreign language, numerous learner corpora have been collected. Here, we will only discuss those that are at least partly available to the general public. A more exhaustive list of learner corpora in many languages is provided on the *Center for English Corpus Linguistics* (CECL) website, from UCLouvain in Belgium. Many learner corpora are also available on the TalkBank online database.

The *Corpus écrit de français langue étrangère* or Lund CEFLE Corpus brings together texts produced by Swedish learners of French, aged between 16 and 19 years with varying skill levels. The texts are compositions of a descriptive or narrative nature, as well as stories created on the basis of images for description. The corpus amounts to approximately 100,000 words but only part is publicly available. This portion of the corpus comprises a longitudinal section, where each learner has produced four texts. Learner levels range from initial to very advanced, with three to four students for each level. The cross-sectional portion of the corpus includes 136 texts written based on the same image description task, by learners from the initial level to the advanced level, and by a control group of native speakers. All of these can be downloaded in text format.

The *Dire autrement* corpus, created in Canada by Marie-Josée Hamel and Jasmina Milicevic, contains texts mainly produced by English-speaking learners. It totals approximately 50,000 words and gathers material from different textual genres, either of a narrative or an argumentative nature. The corpus is available on request from the authors.

The *French Learner Language Spoken Corpora* created by Florence Myles and Rosamund Mitchell brings together seven spoken corpora by French language learners. Six of them were collected at English universities and include English-speaking learners of French, who often studied at university level or during high school. The last one (*Brussels' project*) includes Dutch-speaking learners. Learner levels vary depending on the corpus. Some of the corpora are longitudinal and others are cross-sectional. These corpora can be downloaded or viewed via an online interface. All the corpora have been transcribed in CHAT format and can be explored with the CLAN tool (see section 5.6.3).

The *Phonologie du français contemporain* corpus (Detey *et al.* 2009) includes learners of seven different mother tongues, namely English, German, Dutch, Spanish, Norwegian, Japanese and Cypriot Greek, at different learning levels. The corpus includes recordings of aloud readings, word repetition tasks, structured interviews and free conversations between two students. A portion of the corpus is available to the public and can be consulted via an online interface. Moreover, database access is free of charge, after filling in a declaration of use form.

The *Interfra* corpus created by Inge Bartning and Fanny Forsberg Lundell focuses on Swedish learners of French at different levels. The corpus contains interviews, narrations based on videos, and images. The first part of the corpus includes French learners who have been exposed to the language within the context of schooling. On the one hand, it comprises high school students and, on the other hand, university students from beginner to advanced level. A second portion of the corpus focuses on advanced learners who have all lived in France, for a period ranging from 1–2 years to more than 30 years (see Chapter 3, section 3.3 for a study based on these data). Control groups of native speakers were also recorded. The corpus is fully available to the public and can be viewed free of charge via an online interface.

The *University of West Indies Learner Corpus* or UWI L2 Corpus created by Hughes Peters includes material spoken by adult French learners (16 in total) who were also speakers of English and Jamaican Creole, and who had studied French at university. The corpus contains conversations during spoken exams and in informal contexts, and amounts to approximately 15,000 words, 9,500 of which were produced by learners. The corpus has been transcribed in CHAT format and can be downloaded or viewed online.

5.6. Multilingual corpora including French

Most of the time, comparable corpora are assembled by researchers for the needs of their projects from existing monolingual corpora. However, comparable corpora are sometimes already publicly available. This is the case, for example, for the three corpora of parliamentary debates collected by Truan (2016). This resource contains corpora of three parliamentary debates held between 1998 and 2015 at the House of Commons in England (approximately 190,000 words), in the Bundestag in Germany (approximately 420,000 words) and in the National Assembly in France (approximately 137,000 words). Similarly, the *C-Oral-Rom* corpus contains spoken presentations in the four Latin languages – French, Italian, Spanish and Portuguese – gathered in comparable contexts (Cresti and Moneglia 2005). Each language has approximately 300,000 words. This corpus is available as a CD Rom.

The number of parallel corpora available is constantly increasing. The OPUS database includes many free access parallel corpora, including the Europarl corpus, described below, as well as corpora with subtitles and multilingual data collected from the Internet, such as Wikipedia. These data have been automatically annotated with part-of-speech taggers; however, these annotations have not been verified manually so there remain a small percentage of errors.

The *Multilingual Corpora for Cooperation* (MLCC) project aims to bring together both parallel and comparable corpora. The parallel portion of the corpus, *Multilingual Parallel Corpus*, contains texts translated into nine European languages: German, English, Danish, Spanish, French, Italian, Greek, Dutch and Portuguese. The data in this corpus have been drawn from two sources:

– the *Official Journal of the European Commission* C series, *Written Questions* from 1993, which corresponds to more than 10 million words;

– the *Official Journal of the European Union, Annex: Debates of the European Parliament* 1992–1994 and which amounts to a total of 5–8 million words per language. The comparable portion of the corpus contains articles from financial newspapers of the early 1990s, in six languages: German, English, Spanish, French, Italian and Dutch. This resource is available for free via the ELRA website.

The parallel portion of the MLCC has been rendered somewhat obsolete by the development of the Europarl corpus (Koehn 2005) which contains debates of the European Parliament between 1996 and 2011. The principle adopted in the European Parliament is that each Member has the right to speak in their own language and that speeches should then be translated into all the other languages of the Union. The result of such multilingualism is a huge multilingual corpus. In its current version (version 7), the corpus is distributed as 20 parallel corpora in English and another European Union language.

However, these corpora include not only the speeches produced originally in each language, but also their translations. Given the fact that the Europarl corpus was mainly compiled to serve as training material for machine translation systems, the difference in status between the original and translated languages is of little importance. Nevertheless, to carry out contrastive studies, it is very important to have access to this information (see Chapter 4, section 4.4). This is why Cartoni *et al.* (2013b) compiled real directional corpora based on Europarl data, where source and target languages are clearly identified (see Chapter 4, section 4.5, for an example of how to use these corpora). These directional corpora can be obtained from the authors.

Also drawn from the European Union documents, the *JRC-Acquis* corpus (Steinberger *et al.* 2006) contains all the laws which can be enforced in European Union countries. They are thus legislative texts dating from the 1950s to the present day. The corpus is a collection of parallel texts in 23 languages, comprising a total of more than a billion words. The corpus is aligned per sentence in 231 language pairs, meaning that it is one of the most multilingual resources in the Europarl corpus. However, identifying the status of each text as original or translated language in this corpus is a difficult task. It can be downloaded for free from a research-related European Union website.

The Hansard corpus is also made up of parliamentary debates, more specifically, from the Canadian Parliament. Therefore, debates are in French and English, and accompanied by translations into the other language. This corpus includes spontaneous language, prepared speeches and written texts. The version of the corpus which is available free of charge online includes 1,300,000 aligned sentences or fragments, amounting to approximately 2 million words per language.

Representative of another type of language register, TED conferences have recently made it possible to create a large parallel corpus. Indeed, the presentations in English which can be viewed on the TED website have been transcribed and then translated into many languages (for subtitling purposes) by voluntary users. These translations make it possible to study translation equivalents in a completely different register from the legal and institutional style of the Europarl and Hansard corpora. However, these translations are generally not the product of professional translators, as they are made by volunteers. Taking this into account, the presence of errors should not be discarded, nor should the fact that style may not fully reflect that of professional translations (see Chapter 4, section 4.4, for a study comparing the translations from the Europarl and the TED corpora). Another limitation of this corpus is that it is unidirectional. Actually, TED Talks are always made in English; as a result, English is the only source language, contrary to the Europarl corpus, where all languages are alternately source and target. In the TED Talks corpus, the only variations concern the numerous target languages.

Regarding yet another genre, the CRATER corpus (Garside *et al.* 1994) is a parallel trilingual French, English and Spanish corpus, containing texts in the field of telecommunications. The second version, the CRATER2 corpus, contains one and a half million words in French and English, and 1 million words in Spanish. These corpora are distributed for a fee via the ELRA website.

In the literary field, the *ParCoGLiJe* corpus (Stosic *et al.* 2018) is a parallel English–French corpus which includes children's literary classics in both languages. The corpus consists of four texts in French (*Lettres de mon moulin*, *Les trois mousquetaires*, *Mémoires d'un âne* and *Vingt mille lieues sous les mers*) and four texts in English (*Oliver Twist*, *The Secret Garden*, *The Jungle Book* and *Treasure Island*), and their translations. All of the texts, including the original as well as the translated language, make up a corpus of more than 1,600,000 words. The texts are provided in XML format and have been sentence aligned, making it possible to search through this corpus using a bilingual concordancer. This corpus is available free of charge on the Ortolang platform.

In the area of new media, the Swiss portion of the *sms4science* corpus collected by the universities of Zurich and Neuchâtel (Dürscheid and Stark 2011) includes SMS messages in the Swiss national languages. More specifically, this corpus contains a total of nearly 26,000 text messages produced by approximately 2,800 users. Among these SMS, 41% are in Swiss German, 28% in standard Swiss German (*Hochdeutsch*), 18% in French, 6% in Italian and 4% in Romansh. This corpus is accessible by logging in to the dedicated site. A similar corpus containing WhatsApp exchanges is also distributed.

The English–French *Cabal2* parallel corpus, produced by Poitiers University, at the laboratory *Formes et representation en linguistique et littérature* or FORELL includes journalistic texts, most of which have been drawn from *Le Monde Diplomatique* between 1998 and 2003. The other sources are *Courrier International*, *Time Magazine*, *National Geographic* and some chapters from Jules Verne's novels. In total, this corpus includes 200 articles which correspond to approximately 400,000 words. The corpus can be queried online. The results provide the sentence in which the looked-up word appears, together with its translation. This tool is very useful for quickly finding examples of word translations but it cannot be used to perform a truly quantitative contrastive analysis, as the total number of occurrences of the word is not mentioned, nor is the translation's direction. In addition, the tool is not suitable for complex queries.

5.7. Corpus consultation tools

Concordancers are tools specifically designed for corpus analysis. In this section, we will begin by briefly introducing their main features. We will then focus on the freely available AntConc concordancer, and introduce some of its functionalities. Finally, we will discuss briefly the features of the CLAN concordancer which makes it possible to explore data coded in CHAT format, the annotation standard used in the CHILDES database.

5.7.1. *Concordancers*

Above all, a concordancer is a tool that makes it possible to look up words in their context of use. For instance, in the *Littéracie avancée* corpus described above, a search for the French word *avis* (opinion) by means of a concordancer indicates that the students have used this word 61 times. It also helps visualize the sentences in which it was used, aligned per occurrence of the word retrieved, as we can see in the search results reproduced in Figure 5.1.

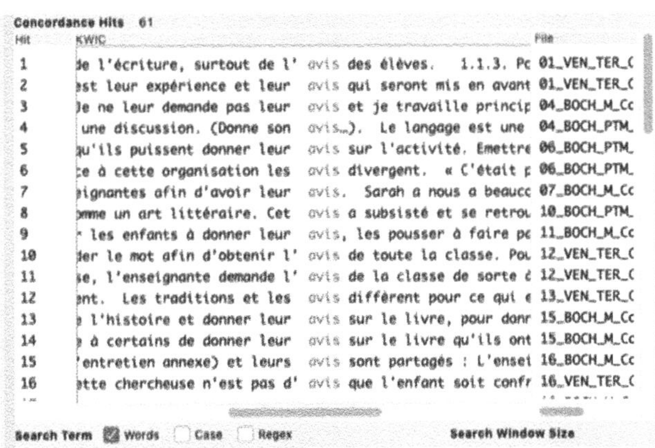

Figure 5.1. *Search results for the word "avis" with AntConc. For a color version of this figure, see www.iste.co.uk/zufferery/corpus.zip*

Thus, searching for a word in a concordancer quickly gives some indication, not only of the frequency of certain words in a corpus, but also of the linguistic context in which they appear. Indeed, a concordancer generally makes it possible to sort word occurrences according to their neighbors to the left or to the right. Here, sorting such as this indicates that the word *avis* often appears in the corpus before words like *et* and *sur,* and to a lesser extent, together with *ainsi* and *car*. It also shows that when the noun *avis* is modified by an adjective, the most frequent adjectives are *différents* and *divergents*.

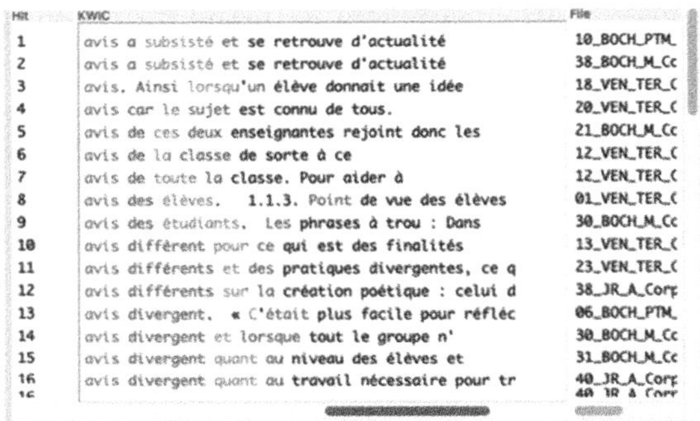

Hit	KWIC	File
1	avis a subsisté et se retrouve d'actualité	10_BOCH_PTM_
2	avis a subsisté et se retrouve d'actualité	38_BOCH_M_Cc
3	avis. Ainsi lorsqu'un élève donnait une idée	18_VEN_TER_(
4	avis car le sujet est connu de tous.	20_VEN_TER_(
5	avis de ces deux enseignantes rejoint donc les	21_BOCH_M_Cc
6	avis de la classe de sorte à ce	12_VEN_TER_(
7	avis de toute la classe. Pour aider à	12_VEN_TER_(
8	avis des élèves. 1.1.3. Point de vue des élèves	01_VEN_TER_(
9	avis des étudiants. Les phrases à trou : Dans	30_BOCH_M_Cc
10	avis diffèrent pour ce qui est des finalités	13_VEN_TER_(
11	avis différents et des pratiques divergentes, ce q	23_VEN_TER_(
12	avis différents sur la création poétique : celui d	38_JR_A_Corp
13	avis divergent. « C'était plus facile pour réfléc	06_BOCH_PTM_
14	avis divergent et lorsque tout le groupe n'	30_BOCH_M_Cc
15	avis divergent quant au niveau des élèves et	31_BOCH_M_Cc
16	avis divergent quant au travail nécessaire pour tr	40_JR_A_Corp

Figure 5.2. *Occurrence sorting of the word "avis" according to its neighbors to the right, using AntConc. For a color version of this figure, see www.iste.co.uk/zufferery/corpus.zip*

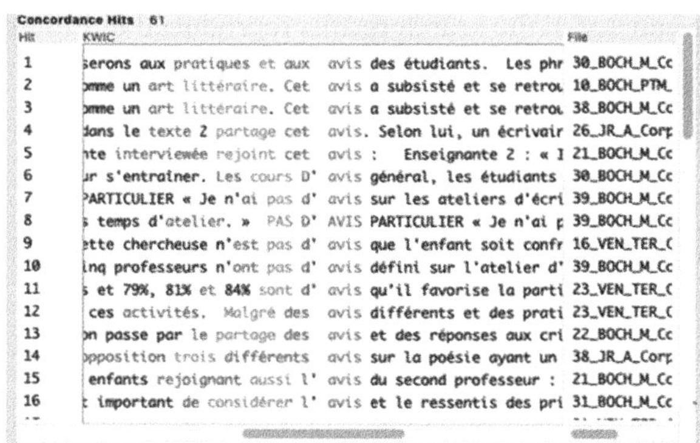

Concordance Hits 61

Hit	KWIC		File
1	serons aux pratiques et aux	avis des étudiants. Les phr	30_BOCH_M_Cc
2	omme un art littéraire. Cet	avis a subsisté et se retrou	10_BOCH_PTM_
3	omme un art littéraire. Cet	avis a subsisté et se retrou	38_BOCH_M_Cc
4	dans le texte 2 partage cet	avis. Selon lui, un écrivair	26_JR_A_Corp
5	nte interviewée rejoint cet	avis : Enseignante 2 : « I	21_BOCH_M_Cc
6	ur s'entrainer. Les cours D'	avis général, les étudiants	30_BOCH_M_Cc
7	PARTICULIER « Je n'ai pas d'	avis sur les ateliers d'écri	39_BOCH_M_Cc
8	s temps d'atelier. » PAS D'	AVIS PARTICULIER « Je n'ai p	39_BOCH_M_Cc
9	ette chercheuse n'est pas d'	avis que l'enfant soit confr	16_VEN_TER_(
10	inq professeurs n'ont pas d'	avis défini sur l'atelier d'	39_BOCH_M_Cc
11	s et 79%, 81% et 84% sont d'	avis qu'il favorise la parti	23_VEN_TER_(
12	ces activités. Malgré des	avis différents et des prati	23_VEN_TER_(
13	on passe par le partage des	avis et des réponses aux cri	22_BOCH_M_Cc
14	opposition trois différents	avis sur la poésie ayant un	38_JR_A_Corp
15	enfants rejoignant aussi l'	avis du second professeur :	21_BOCH_M_Cc
16	t important de considérer l'	avis et le ressentis des pri	31_BOCH_M_Cc

Figure 5.3. *Occurrence sorting of the word "avis" according to its neighbors to the left, using AntConc. We can see that the noun "avis" is often preceded by pronouns "cet" (this) and prepositions like "d'" (of). A word further to the left, we find verbs like "partager" (share), "rejoindre" (join) and "considerer" (consider). For a color version of this figure, see www.iste.co.uk/zufferery/corpus.zip*

In order to analyze the typical environment of a word, concordancers help determine which words co-occur most frequently with the word looked-up. Although the list provided above gives us an approximate idea of the frequent co-occurrences, it does not let us quantify such associations. For example, in the *Littéracie avancée* corpus, the five most frequent co-occurrences to the right of the word *avis* are *avis sur*, which appears 11 times and then *avis et* (six times), *avis de* (five times), *avis divergent* (four times) and *avis des* (three times). This list also makes it possible to identify other modifiers of the noun *avis* apart from *divergent*, on the basis of the observation of the list of occurrences. Frequent modifiers are *différent*, *particulier*, *défini*, *général*, *mitigé*, *personnel* and *respectif*. The search for co-occurrences to the left indicates that the most frequent elements found in the corpus are: *leur avis* (18 times), *les avis* (seven times), *d'avis* (six times), *l'avis* (six times) and *son avis* (five times). Some concordancers can calculate the probabilities of collocations between certain words, rather than simply establishing the list of words which co-occur in the corpus. In the case of the word *avis*, the most likely collocations calculated by Antconc are: *subsistés*, *respectifs*, *émettre*, *défavorable*, *réponses* and *divergentes*.

Finally, some concordancers can be used to extract a list of keywords in a corpus by comparing them with a reference corpus (see Chapter 6). More specifically, the concordancer determines which of the words are used significantly more (or even less used in the case of negative keywords) in the search corpus compared to the reference corpus. Another possibility is to take a portion of a corpus and compare it to the rest of the corpus, or to compare a corpus with another similar corpus. For example, in the *Littéracie avancée* corpus, we can identify the keywords which specifically match student reports, compared with other types of academic work such as dissertations, by comparing this sub-section of the corpus to the others. The resulting list of keywords includes common nouns such as *réflexivité*, *résumé*, *généralisation*, *portfolio*, *globalisation*, *article*, *stagiaires*, etc., as well as proper nouns like *Salaün*. The presence of proper nouns in the keyword lists is very frequent because these words often specifically refer to a particular person, which is not used equally in different corpora. In the case of the noun *Salaün*, its presence in the keywords of the corpus can be explained by the fact that Salaün was a general delegate of a road prevention association who had taken part in an interview that students had to discuss in one of their assignments.

Setting up a list of keywords also makes it possible to compare the themes of different works by the same author. For example, if we compare the novel by Jules Verne *Le Tour du monde en 80 jours* with his other novel *Vingt Mille Lieues sous les mers,* the specific keywords of the first one are common nouns such as *train, gentlemen, voyageurs* and *paquebot*, as well as proper nouns like *Fogg*, *Passepartout*, *Phileas*, *Hong Kong*, etc. In another register, a comparison of the *Le Monde* corpus from the year 2011 with that of the year 1987 generates keywords in the form of common nouns such as *euros*, *economy, Internet* and *international* and proper nouns like *Sarkozy*, *Hollande*, *Obama*, *Aubry* and *Merkel*. This list is a good reflection of the important topics and personalities discussed in 2011 whom we did not yet talk about in 1987. When we compare the years 1987–1995 of the *Le Monde* newspaper, keywords change to: *serbes*, *ETA*, *Bosnie*, *Sarajevo*, *Croatie*, *Jospin*, *Balladur*, etc.

In a nutshell, the keyword list of a corpus is very useful to identify its main topics, provided that the comparison with the reference corpus is appropriate. Indeed, the latter is of paramount importance in establishing the list of keywords. If we compare the reports of university students with a year of the *Le Monde* newspaper rather than the rest of their academic work, the keyword list includes *élèves*, *activité*, *formation*, *réflexivité*, *écriture*, *évaluation*, *résumé*, *pensée*, *enseignants*, *savoir*, etc., because the topics covered vary more widely between the two corpora than between the different types of university work. This is why the topics emerging from this second comparison are those related to the field of education in general, rather than those addressed in the reports in particular.

In sum, concordancers make it possible to analyze recurrent properties in a corpus, such as its frequent words, its collocations and its keywords from a quantitative point of view, something which is not possible to infer from simply reading texts. This is why they represent essential tools for grasping the quantitative properties of a corpus.

5.7.2. *Focus on the AntConc concordancer*

The AntConc concordancer, developed by Laurence Anthony, is available for free online. AntConc can be used to perform all the analyses described above. This concordancer is compatible with the various current operating systems (Linux, MacOS or Windows). In this section, we will describe its basic principles of use. For a start, AntConc can only read text format files. So, to begin with, it is necessary to convert the files included in the corpus to text format. Depending on their origin, this can be rather easy or more difficult: usual word processors generally have a text format saving functionality, but extracting text from a PDF file can be difficult. AntConc can also read XML files, since these contain text which is accompanied by tags. Before opening one or more files for them to be processed with AntConc, we have to make sure that the encoding chosen in AntConc for reading the characters is suitable for reading the file correctly. By default, AntConc uses UTF-8 encoding. However, this encoding does not correspond to text files containing French characters, because of accented characters. When the text of a file is not displayed correctly in AntConc, the encoding used can be changed to make it match the file's encoding, for example, the ISO 8859-1 format (Latin.1).

AntConc can be used for looking up words in context and sorting their occurrences depending on the words that appear to the left or to the right of the search word. To look up certain words, the use of wildcards can be of great help. All the wildcards recognized by AntConc for defining the search pattern can be easily viewed in the software. These wildcards are mainly used for looking up all the possible endings of a regular verb in a single request, by searching for the radical of the verb followed by any number of characters (through the use of an asterisk), such as *donn**. To look up a singular and a plural word in a single query, it is possible to replace zero or one character exactly with another wildcard, for example, *homme+*.

AntConc also offers the possibility of visualizing the places where various occurrences can be found in a file by means of a graph, as well as going through the entire file until we find where the occurrence originated. This functionality is very useful to obtain the maximum amount of contextual information and thus disambiguate certain occurrences. The Clusters and Collocates tabs help you to identify the collocations spotted in the corpus, as well as the most likely collocations.

AntConc has another feature which offers the possibility of generating a list of all the words in the corpus sorted by frequency via the Word List tab. This same tab also shows the number of word types and word occurrences in the corpus (see Chapter 8, section 8.2, for a discussion of these concepts).

These figures are essential for performing lexical diversity calculations on corpus data, such as the type/token ratio (see Chapter 8). Finally, AntConc makes it possible to create a list of keywords from the corpus based on the comparison with a reference corpus.

5.7.3. *Focus on the CLAN concordancer*

The CLAN concordancer works on files encoded in CHAT format, which corresponds to all of the data in the CHILDES database, as well as a number of learner corpora. CLAN can be installed on Mac and Windows operating systems. CLAN commands can also be used with the online version of the corpora.

CLAN offers the possibility of formulating queries on the CHILDES corpus for studying many aspects of children's language. Although query syntax may seem complex at first, it is actually easy to master. A request in CLAN should always start by specifying the name of the command to be performed, followed by the search elements, the file or file's name where the information should be retrieved from, and whether it should be related to the whole corpus or only narrowed to some files. Finally, if applicable, the command should specify the type of speaker whose words have to be analyzed. This specification is often very useful, since the interactions in acquisition corpora most often take place between children and one or more adults and it is necessary to analyze the speech produced by each of them separately.

One of the most useful CLAN commands is the *combo* command, which helps you to look up words or word sequences produced by specific speakers in the corpus. If the corpus has been annotated, this command also makes it possible to search for grammatical categories, speech acts or even errors. In CHAT format, annotations always take the form of an additional line below the transcription, identified as %mor, for example, when referring to a grammatical category or the morphological representation of a word. The coding of speech acts is identified with a line called %spa. Relevant nonverbal actions are coded with a line called %act. Finally, error coding is identified with a line called %err and has a standardized format. For instance, the $LEX reference indicates a lexical error. In the transcription itself, incorrect words are followed by an asterisk in square brackets so that they can be identified. Let us take a look at an utterance from the York corpus (De Cat and Plunkett 2002):

> *CHI:tu me l'as donné.
> %mor:pro:subj|tupro:obj|me
> pro:subj|il$v:aux|avoir&PRES&2spart|donner-PP&m.
> % act: takes a book

Going back to the *combo* command, to find all the occurrences of the word *pourquoi* produced by the child, the syntax of the command should be formulated as follows:

```
combo +spourquoi +t*CHI
```

CLAN also helps you to determine the frequency of words in a CHAT file using the *freq* command. This command makes it possible to obtain the list of words sorted by frequency, in the same way as the list of words generated by AntConc. The command also helps you to calculate the type/token ratio (see Chapter 8), which represents a measure of lexical diversity. The syntax for such a command is as follows:

```
freqFILE NAME +t*CHI +o
```

Finally, the complexity of children's language is often measured at the start of their development by their mean length of utterance (MLU) (see Chapter 3, section 3.1). The MLU can be calculated automatically in CLAN using the MLU command. To do this, the syntax is very similar to the other commands:

```
mluFILE NAME +t*CHI
```

5.8. Conclusion

In this chapter, we have presented the main corpora available in French. We have observed that, despite the absence of a reference corpus, numerous more specific corpora are available, which can be combined to carry out research in many areas of linguistics, as we will see in the subsequent exercises offered.

The main limitation for using these corpora is their availability, which is often limited and requires going through an online interface, in which only some functionalities can be used. When corpora can be downloaded, a concordancer should be used in order to explore them systematically. Finally, we reviewed the main features of concordancers and presented two of them succinctly: AntConc and CLAN.

5.9. Revision questions and answer key

5.9.1. *Questions*

1) Using the interface provided on the website of the *Corpus français de l'université de Leipzig*, what is the frequency order in this corpus of the words *maison*, *chalet*, *immeuble*, *bungalow*. What are their most frequent collocations? What can you conclude from these observations?

2) Using the AntConc concordancer, find the 10 most frequent content words (defined as nouns, lexical verbs, adjectives and adverbs) used by the undergraduate students in the *Littéracie avancée* corpus. What can you conclude? In this same corpus, what are the five most frequently observed co-occurrences and the five most probable collocations for the word *élève(s)*?

3) Using the *Google Books* corpus online interface, find:

a) when the new spelling of the word *clé* started replacing the old spelling *clef*;

b) which researcher is more popular, Ferdinand de Saussure or Noam Chomsky;

c) whether the nominal use of the word *orange* preceded or followed its adjectival use in the history of French.

4) **Use the online interfaces of the OFROM corpus and the CFPQ corpus**. How often do men and women use the verb *détester* in French-speaking Switzerland and Quebec? What remarks can you make about the possibilities offered by these interfaces?

5) In the York language acquisition corpus on the CHILDES database, what is the most frequent word produced by Anne in the first recording at the age of 1 year and 10 months old? What about the last recording, at 3 years and 5 months old? How did its type/token ratio change in these two

files, and what is the MLU in both files? Compare with the results for Max in the same corpus. Based on these clues, who seemed to acquire language the fastest?

6) From the TED corpus, identify the different possible translations of the word *issue* into French.

5.9.2. *Answer key*

1) The word *maison* has 282,802 occurrences in the corpus. It is the 474th most frequent word and takes frequency class number 8. It is the most frequent word in the group. The word *immeuble* is the second most frequent word, with 20,133 occurrences, making it the 6,633th most frequent word in the corpus and corresponding to frequency class number 12. The third most frequent word is *chalet*, with 8,184 occurrences in the corpus, corresponding to frequency rank number 13,609 and frequency class number 13. The word *bungalow* is the least frequent word, with 1,100 occurrences in the corpus, corresponding to frequency rank number 53,542 and frequency class number 16. As shown in the graph provided on the website, the main collocations of *maison* are the nouns *mère*, *retraite*, *disque*, *famile* and *jardin*, the adjective *familiale* and prepositional phrases such as *d'édition* and *d'arrêt*. The collocations for the word *immeuble* are the nouns *quartier*, *logement*, *appartement*, *étages*, *incendie* and *bureaux,* the prepositional phrase *d'habitation*, the past participle *situé* and the demonstrative *cet*. The collocations for the word *chalet* are the nouns *bois*, *ski*, *montagne*, *location*, *résidence* and *vacances*, the prepositional phrase *d'alpage*, the past participles *situé* and *assigné*, as well as the proper nouns Roman Polanski and Gstaad. Finally, the collocations for the word *bungalow* are the nouns *plage*, *villa*, *location*, *chambre*, *chalet*, *vacances*, *camping*, *maison*, *mobil-home*, as well as the adjective *petit* and the prepositional phrase *d'accessibilité*. We can see from this list that the meaning of the words can, at least in part, be inferred from their collocations. We can also observe that the word *maison* is the most generic of the four, and the only one that takes figurative meanings as in *maison d'édition, maison d'arrêt* or *maison de disques*. When compared with *immeuble*, we can see that the word *maison* also has its own attributes, like *jardin* and *famille*. Conversely, *maison* is associated with *appartements* and *étages*, which specifies the type of housing in question, as well as *bureaux* which shows a different use from that of *maison*. Finally, *chalet* and *bungalow* are both associated with

vacation homes, but of a different kind. While *chalet* is associated with *ski*, *montagne* and *bois*, *bungalow* is associated with *plage* and *camping*. The proper nouns associated with *chalet* in the corpus show one of the limitations of the collocation analysis. Indeed, in the corpus, several articles referred to Roman Polanski's residence, which made these associations very strong, but these words do not obviously collocate in everyday language.

2) In order to answer this question, it is necessary to open AntConc, and there, to open the files in the L2_DOS_SORB and L3_RS_BOCH directories, which correspond to material by undergraduate students. Then, we have to generate the word list under the Word List tab. The 10 most frequent content words are the following:

Rank	Number of occurrences	Word
32	499	*plus*
33	469	*enfants*
38	423	*élèves*
47	291	*fait*
49	279	*tout*
50	264	*meme*
51	259	*faire*
52	258	*classe*
53	255	*deux*
59	218	*travail*

This list illustrates the fact that the most frequent words in a corpus are those belonging to functional categories such as prepositions and determiners. Indeed, the first content word only appears at the 32nd frequency rank! We can also observe that the frequency of words in a corpus decreases rapidly. The most frequent word in the corpus, that is, *de*, has 4,461 occurrences, whereas the 32nd word has only 499 occurrences, representing almost 10 times fewer occurrences. In addition, from frequency rank number 4,077 onwards, the words in the corpus only have one occurrence. This distribution reflects Zipf's law (see Chapter 6). The five most frequent co-occurrences to the right of the word *élève*(*s*) are: *de*, *ont*, *et*, *en* and *avaient*. To the left, these are the words: *les*, *des*, *l'*, *aux* and *un*. The five most likely collocations are: *répartie*, *accompagné*, *onze*, *évaluerai* and *équitablement*.

3) a) To answer this question, we have to type "*clé, clef*" on the online corpus interface. We should also be careful to choose the French corpus and to determine a sufficiently long time period, for example from 1800 to 2000.

The results obtained indicate that the spelling *clé* became as frequent as *clef* in 1963 and has made strong progress since then, to the detriment of the old spelling.

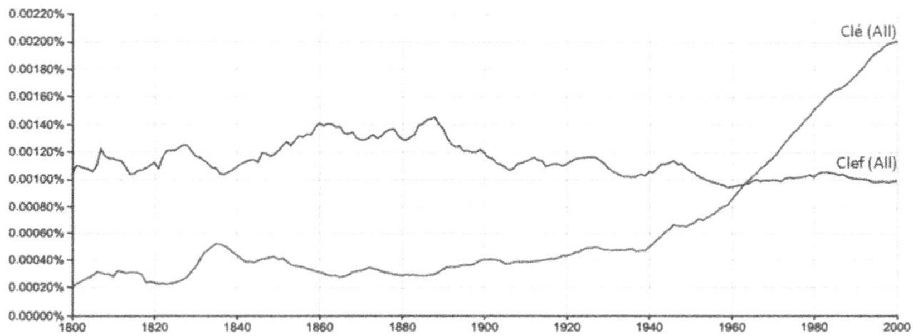

Figure 5.4. *Frequency of the words* clé *and* clef *from 1800 to 2000 in the Google books corpus. For a color version of this figure, see www.iste.co.uk/zufferery/corpus.zip*

b) By searching for "Saussure, Chomsky" in the French corpus of the 20th Century, we can note that Saussure has always been, and remains, a more popular linguist than Chomsky in the French-speaking literature. In addition, while the popularity of Chomsky has tended to decrease, that of Saussure has remained stable since the 1970s.

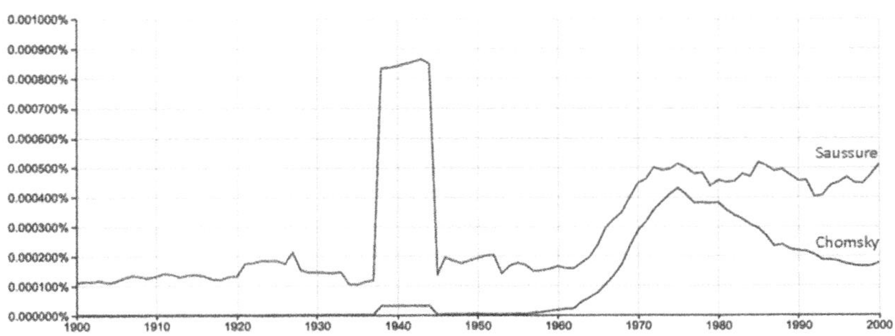

Figure 5.5. *Frequency of the words Saussure and Chomsky from 1900 to 2000 in French in the Google books corpus. For a color version of this figure, see www.iste.co.uk/zufferery/corpus.zip*

On the other hand, the same research in the English corpus indicates that, in the English-speaking culture, Chomsky replaced Saussure in the early 1960s. This corpus also indicates that the popularity of Chomsky has been declining since the 1980s.

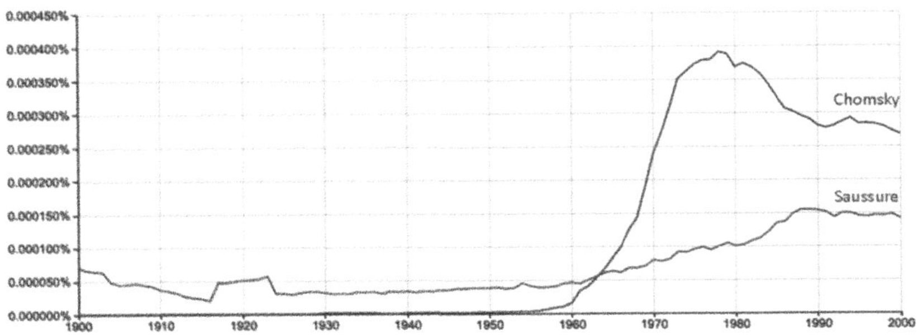

Figure 5.6. *Frequency of the words Saussure and Chomsky from 1900 to 2000 in English in the Google books corpus. For a color version of this figure, see www.iste.co.uk/zufferery/corpus.zip*

c) For this search, a very long time interval should be specified, for example, between 1600 and 2000. To differentiate the two uses, we should look up each grammatical category separately by specifying "orange_NOUN, orange_ ADJ". We can observe that the noun *orange* appeared for the first time in the corpus in 1624, whereas the adjective frequency only increased from the middle of the 18th Century onwards.

Figure 5.7. *Frequency of the word orange as a noun and as an adjective from 1600 to 2000 in French in the Google books corpus. For a color version of this figure, see www.iste.co.uk/zufferery/corpus.zip*

4) In the French corpus of French-speaking Switzerland, the verb *détester* is used five times by men and 15 times by women. This search can be done very easily by looking for the infinitive *détester* in its lemmatized form, which helps us to find all the inflected forms in a single request. In the corpus of spoken French from Quebec, there is no occurrence of the verb *détester* produced by men, versus 16 occurrences produced by women. The search is much more complicated in this interface, since the search by means of lemmatized forms is not possible. All verb forms must be looked up separately.

5) The most frequent word produced by Anne at 1 year and 10 months old was *no*, with 32 occurrences. At this age, her type/token ratio was 0.37. At 3 years and 5 months old, the most frequent word was *ça* with 54 occurrences, and her type/token ratio was 0.21. In addition, her MLU increased from 1.82 to 4.83 from the first to the last recordings. The most frequent word produced by Max at the start of the corpus at 1 year and 9 months old was also *no*, with 24 occurrences. At this age, his type/token ratio was 0.38. At 3 years and 2 months old, at the end of the corpus, the most frequent word was *I*, with 48 occurrences. The type/token ratio was 0.24. His MLU ranged from 1.17 at the start of the corpus to 3.78 at the end. The comparison of MLU between Anne and Max seems to indicate that Anne developed her language faster than Max. The fact that the type/token ratio decreases with age in the two children reflects that the total number of occurrences they produce increases a lot as recordings progress (e.g. ranging from 298 occurrences to 1,092 occurrences for Anne), which implies a poorer lexical diversity in proportion to the total number of words produced. The type/token ratio cannot therefore be considered as a reliable measure of linguistic development. A better appraisal can be obtained by comparing the number of different words, known as word types, produced by each child. At the end of the corpus, Anne produced 230 word types against 182 for Max, which tends to confirm that her language was more advanced.

6) The English noun *issue* was used 1,957 times in the TED conference corpus. It is therefore difficult to study all of these occurrences. To quickly determine frequent translations, 100 occurrences can be randomly chosen. This observation of translations gave the following French translations for the first 100 occurrences of the word *issue*:

Number of occurrences	Translation
57	*problem*
16	nothing or paraphrase
9	*question*
7	*enjeu*
6	*sujet*
2	*cas*
2	*affaire*
1	*difficulté*

We can observe that the main translation of the noun *issue* in to French was *problème*. We can also reason that the word *issue* does not always have an exact equivalent in French, which might explain the high number of untranslated occurrences, or translations by means of a paraphrase. In particular, this word is used in expressions such as *this issue,* often translated by pronouns like *cela* in French. This research project also revealed the difficulties inherent in the observation of translations. In fact, the TED interface does not currently let us specify the search for words or character strings. Thus, the search also generates irrelevant occurrences of words like *tissue*, as well as many cases of sentences which have not been translated into French. For this search, it was necessary to consider 175 English sentences in order to find 100 translations of the noun *issue* into French.

5.10. Further reading

A list of existing corpora in many different languages can be found in Ostler (2009). A list of corpora more specifically devoted to English is available in Lee (2010). For the French language, the Ortolang platform contains a broad choice in corpora, from which we only presented a portion in this chapter. A list of learner corpora around the world can be found on the CECL group website from UCLouvain. The functions of concordancers have been described by Evison (2010). A very accessible online tutorial regarding the AntConc concordancer can be found on the Programming Historian website, and the functionality of the CLAN concordancer is described in detail by MacWhinney (2000, pp. 130–279). For all the sites mentioned, refer to the webography at the end of this book.

6

How to Build a Corpus

In this chapter, we will present the best practices for creating a corpus. First, we will discuss some facts that need to be considered before deciding to create a new corpus and highlight the advantages of reusing existing data whenever possible. Then, we will address various important methodological concerns for creating a corpus, in particular questions related to the size and representativeness of samples, and will explain simple methods for data sampling and coding. We will also briefly discuss the challenges posed by the creation of the spoken corpora. We will finally see that the task of creating a corpus carries with it a certain number of ethical and legal issues which must be dealt with.

6.1. Before deciding to build a corpus

The first element to check before starting to compile a new corpus is whether existing data can be used for the planned study. As we will see throughout the chapter, creating a corpus is a challenging task and presents many difficulties. It is actually not always easy to find texts available in a digitized format for all text genres, and even when such texts exist, they might not all be usable due to copyright issues. Choosing the right texts to be included in a corpus should also be the object of careful reflection, since any kind of analysis carried out on data that are not representative of the target genre (see section 6.2) could be largely invalid. When it comes to creating a reference corpus, the data collection phase is so time-consuming that it can only be tackled by a group of experts. Becoming involved in a corpus creation project individually is realistic only in the case of specialized corpora, for example, if the task is narrowed to a specific language register or a regional variety, that is, a project of a smaller size. Even for this type of corpus, several months of work are often necessary for collecting the data, and may take even longer if the latter are enriched with linguistic annotations (see Chapter 7).

The problems are even more complex and numerous when it comes to spoken data. These need to be collected in the form of audio files which are later transcribed to become analyzable with corpus searching tools. The transcription process itself is very time-consuming and its complexity depends on the exact type of annotation that is added to the data (prosodic contours, etc.). To get an idea of the magnitude of the task, up to 15 hours of work are necessary to transcribe one hour of recording (Reppen 2010b, p. 34). Transcription also poses methodological challenges (see section 6.5), for example: how should we annotate hesitations, false starts and variations in pronunciation? How should we account for the overlaps in speech turns in conversations? In addition, the use of spoken data often requires aligning the transcription with the sound file, so as to offer users the possibility of listening to excerpts from the corpus. This sound/text alignment process requires the technical know-how which can be difficult to acquire for inexperienced researchers. For all these reasons, it is preferable to reuse existing data whenever possible.

In Chapter 5, we saw that many corpora in French have already been created, and that some of them are available free of charge to the public. Some other European languages, not only English but also German, Dutch, Spanish and others, have an even broader choice of corpora than French. So, when formulating an empirical research question, it is advisable to consider whether these resources could not be used for the study. If necessary, existing data can be supplemented with a smaller portion of new data, and thus significantly simplify the data collection phase. For example, an empirical study on the regional differences in the way questions are formulated in spoken French could reuse data collected from different spoken corpora, including France, French-speaking Switzerland, Quebec and Belgium. If the study were to be extended to other regional French varieties, for example, the French spoken in the Caribbean islands, existing data could be supplemented by samples of such variety. In Chapter 4, we also saw that comparable corpora can often be assembled from existing data. For example, Crible (2018) studied how discourse markers like *bon*, *ben* and *voilà* in spoken French and *well*, *I mean*, *you know* in spoken English are used in eight different spoken registers which vary depending on certain parameters, such as the degree of pre-planning and whether they were dialogues or monologues. In order to be able to work with comparable multilingual corpora in each speech genre, Crible reused existing data. In English, she used a British portion of the *International Corpus of English* (ICE-GB) and in French, due to the absence of a reference corpus, she assembled a corpus from a series of existing spoken (transcribed) corpora, such as the Valibel database and the *CLAPI* corpus (see Chapter 5).

If, after research, it turns out that the existing corpora are not suitable, then the creation of a new corpus might be considered. In this case, it is essential to properly outline the research question that will be studied on the basis of new data, since the latter will have a crucial influence on the whole process, both during the data collection and the annotation phases. In the field of corpus linguistics, it is very common to hear that there are no good or bad corpora, rather there is only corpora which are more or less suitable to address a certain research question. For example, for investigating the expression of subjectivity in journalistic discourse, a corpus entirely made up of editorials would not be appropriate, since this is only a sub-section of the genre, which incidentally is more likely to contain markers of subjectivity than other sub-genres, as dispatches for instance. In this case, two scenarios are possible: either the conclusions of the study will be limited to the editorial style, or the corpus should be diversified in view of including other types of journalistic texts. The problem we have just mentioned involves a key methodological point for corpus studies, which is the representativeness of data. We will discuss this point in the next section.

6.2. Establishing the size and representativeness of data

Let us begin by repeating that there is no ideal size for a corpus, in the same way as there are no intrinsically good or bad corpora. Suffice it to say that the characteristics of a corpus may be more or less appropriate for answering a research question. As the technical capacities of computers have evolved, it has become possible to collect ever larger corpora. Currently, some corpora such as the *Google Books* corpus (see Chapter 5) and the *FrenchWeb 2012* corpus (available on *Sketch Engine*), collected from the Internet, contain several billion words. For a long time, the rule of thumb for collecting a corpus was that it should be as large as possible. The logic behind this principle was that the larger the corpus, the more likely it would contain occurrences of rare linguistic phenomena. Indeed, when the words of a corpus are listed following their frequency order, as in Table 6.1 for the *Sciences Humaines* corpus (available on Ortolang, see Chapter 5), we can observe that the frequency of words decreases very quickly from the beginning of the list.

de (9,174)	et (3,829)	en (2,495)
la (6,405)	le (3,707)	un (2,468)
l' (4,217)	à (3,429)	une (2,455)
des (4,123)	d' (2,586)	du (2,202)
les (4,036)	est (2,496)	que (2,122)

Table 6.1. *List of the 15 most frequent words in the* Sciences Humaines *corpus*

Frequency distribution follows Zipf's law (1932), according to which the most frequent word in the corpus is approximately twice as frequent as the second one, three times more frequent than the third, and so on. The word frequency indicated in Table 6.1 does not follow this decrease exactly, because in French the most frequent words take up different morphological forms. The frequency of lemmatized words is closer to the curve predicted by Zipf. In any case, word frequency declines very quickly in any corpus and many words appear only once. For example, the 100th most frequent word in the *Sciences Humaines* corpus, which is the word *été*, only appears 195 times, whereas the 1,000th word, *devenir*, appears 21 times. Out of the 15,617 different words in the corpus, from the 8,355th position onwards, words only have one occurrence, meaning that almost half of the words in the corpus only appear once. Technically, these are called "hapax words" or *hapax legomena* which in Greek means "mentioned once". We can infer that, due to this distribution, the study of rare words requires the use of large corpora in order to be able to analyze multiple occurrences.

However, more recently, some researchers have defended the idea that maximum size should not always be the goal in the creation of a corpus, since smaller-sized corpora may prove to be adequate for many research questions which do not involve rare words, as we will see in this chapter. In fact, a large corpus is not always suitable for addressing all kinds of research questions. The question of the optimal size for a corpus primarily depends on the nature of the linguistic phenomenon to be studied. The more frequent a linguistic phenomenon, the better it can be studied on the basis of a small corpus. Rarer linguistic phenomena, on the other hand, require larger corpora. This question is also related to the degree of generalization targeted. A corpus representing a specific genre can be relatively small, whereas general corpora need to be much larger.

As we discussed in Chapter 1, a corpus does not simply represent a collection of randomly chosen texts. In fact, a corpus is a collection of texts or recordings specifically chosen in order to be representative of a language, of a certain register or even a language variety. The question of representativeness is therefore essential so that a corpus can be used for answering a research question. In order to fully understand what this notion represents, we will draw an analogy with opinion polls. Let us imagine that we wish to find out which candidate is more likely to be elected in the next presidential elections. In order to find out, it is not possible to ask all the citizens who they intend to vote for. It is therefore necessary to prepare a sample of the population of a more modest size, to whom it might be possible to ask such a question. Later, the results obtained on the basis of this sample can be extrapolated to the entire population. But in order for this technique to work, it is crucial to carefully choose the sample of respondents, in such a way that it represents the whole population. For example, if the sample chosen includes 500 students met at the exit of a university building, the sample obtained will most likely not correspond to the actual result of the election, since this sample is not representative. In fact, students represent only a small portion of the population. In order for the sample to be representative, it should also include people with other types of occupations, different age ranges and from different regions. The same applies to the compilation of a corpus. In order to be a representative, a reference corpus should contain a balanced set of samples covering the main stylistic genres, both in the spoken and written modes. The main issue is to determine the criteria according to which it is advisable to classify the elements included in the corpus to ensure its representativeness. We can be sure about one thing: these criteria should not be related to the linguistic content of the samples, but rather to a classification made on the basis of external criteria, such as text genres and language registers. For example, it would be rather inappropriate to try to study the production of speech acts in the legal context by choosing a corpus exclusively based on a number of performative verbs such as *demand*, *condemn*, *order* that it contains, since this criterion would influence the results found in the analysis afterwards. However, this study would require the assembly of a corpus which tangibly represents the legal language, such as court decisions, because these writings properly match the targeted field of study, that is, the legal language. To sum up, Biber (1993) argued that a selection based on linguistic criteria specifically related to the content of the corpora would drive the analysis work in the direction of a circular path. Indeed, corpora should be used for analyzing the words and the linguistic structures they contain, among other things. If these parameters have been predetermined during the corpus assembling phase, this analysis no longer makes sense.

In the case of reference corpora, sample distribution between different genres is a complex problem in itself, due to the lack of an existing typology of spoken and written genres that is unanimously accepted. To simplify, let us say that written corpora should contain both public texts (published works) and private ones (letters, emails, etc.), collected from different fields such as the press, the sciences or the literature. The spoken section of a corpus should reconcile a variety of choices. It should include both planned and spontaneous spoken speech, monologues and dialogues, drawn from contexts with various degrees of formality. Very often, the creators of new corpora solve the problem of representativeness by following the criteria used in existing reference corpora, such as the *British National Corpus*, a pioneer of the genre.

In cases where researchers need to compile specialized corpora, the question of representativeness is posed a little differently. To continue developing the analogy with polls, if the goal is to know who students will vote for in the presidential elections, it will be enough to interview a sample of students, since such a sample is representative of that population. In the same way, the question of the representativeness of specialized corpora is clearly simplified, because this can be achieved by choosing texts or recordings belonging to a specific genre. However, we should keep in mind that there may be sub-genres within a genre, such as novels, short stories or children's stories within the literary genre, and that these may vary from each other.

Even when working within a text genre, we should aim to diversify its sources as much as possible. For a literary corpus, for example, works from different authors should be included.

From a lexical perspective, the representativeness of data in specialized corpora, such as corpora devoted to newspaper or legal articles, can be measured using the concept of saturation (Belica 1996). This notion means that a certain lexical trait varies very little throughout the corpus. In order to measure saturation, the corpus should be divided into several segments of equal size. A corpus is saturated at the lexical level when the addition of a new segment results in approximately the same number of new words as the previous segment. The usefulness of this measure is nonetheless limited, since it only provides information about the lexical diversity of a corpus, but not about other domains of language.

As a matter of fact, the representativeness of a corpus cannot be ascertained once and for all. In the case of closed corpora (see Chapter 1), data aging implies that they are no longer representative of the most recent developments in the language. In the case of monitor corpora, while the new data added at regular intervals may improve their representativeness, they nonetheless pose balancing problems between the different portions of the corpus, an aspect we will discuss in the following section.

In summary, McEnery *et al.* (2005) are totally right when they affirm that the representativeness of a corpus is more a profession of faith than a scientific reality. From a factual point of view, representativeness cannot be taken for granted. What should really be kept in mind though is the need to build a corpus that best reflects the linguistic style to be studied based on available data, in order to be able to draw appropriate conclusions.

6.3. Choosing language samples

To achieve the representativeness aim discussed above, a corpus should include a sampling of different types of texts or recordings. Unless we are working on a very specific corpus like the Bible or the complete works of an author, most of the time, it is actually impossible to include all the texts or all the recordings belonging to the genre to be studied in a corpus. This is why it is necessary to prepare samples which, once assembled, can work as a representative sub-section of the genre to be studied. The preparation of samples to be included in the corpus poses two important methodological questions: on the one hand, the appropriate size for each sample, and, on the other hand, how to balance the portions of the corpus in such a way that the result is truly representative of the genre.

In order to understand the difficulties of corpus balancing, we will give an example. To be representative, a spoken French corpus should include speakers from different regions, different ages, both male and female. If 95% of the corpus is made up of Parisian speakers aged between 20 and 30 years, such a corpus will not be a representative sample of the population, even if this means including the other speakers among the remaining 5%. While it is true that a sample is expected to bring together a rather restricted version of the overall population, it should also reproduce its main features so that the results obtained on this sample can be extended to the entire population. A corpus of spoken French should therefore include a similar proportion of male/female speakers, from different age groups and different regions. Many other criteria could be included in this selection, such as the socio-economic level of the participants, for instance. As with the question of corpus size, the balancing criterion largely depends on the question the corpus will help to study. In general, a corpus should not be used for establishing a contrast between elements which have not been balanced during the corpus compilation phase. For example, a corpus created for studying the pronunciation of vowels in Paris and Marseille and which has not been compiled representing a balanced sample of different age groups cannot be used carelessly for studying the evolution in the pronunciation of vowels between generations.

In the case of written language general corpora, it is important for the chosen samples to represent different genres, including both published and unpublished texts. In the case of the *British National Corpus* (Aston and Burnard 1998), an English reference corpus, the written texts included were chosen according to three criteria:

– the field, that is, the topic explored in the text;

– the time when the text was produced;

– the distribution mode, depending on whether it was a book, a newspaper or an unpublished text.

The spoken samples were chosen on the basis of demographic criteria such as age, gender, geographic region and social class, as well as contextual criteria. However, the creators of large corpora have agreed that it is very difficult to fulfill all the criteria to achieve a perfectly balanced corpus. One of the major problems is the difficulty of incorporating new data, an aspect which tends to create bias around the choice in favor of more readily available data. Such difficulty is largely due to copyright issues, which we will address in section 6.6. This issue prevents the inclusion of recently published texts in a corpus that have not yet fallen into the public domain. In addition, published texts are generally more easily accessible than unpublished texts, such as emails or personal letters. Finally, texts published on the Internet are much easier to access than texts published on paper. These differences inevitably induce a certain bias towards specific text categories. In the end, balancing a corpus is never a perfect task. As Nelson (2010, p. 60) pointed out, the end result of a corpus is always a compromise between the desires of its creators and the data that can be obtained.

In concrete terms, balancing the portions of a corpus can be achieved by defining a sampling frame which delimits the population to be sampled and lists its relevant properties. Each of these properties should then be mirrored by the corpus sample proportionally to its prevalence in the population. For example, in order to create a corpus of French, speakers from different regions should be included in the sample. Therefore, geographic region becomes a relevant trait for the sampling frame of a spoken French corpus. The number of French speakers to be included for each region can be determined proportionally to the number of French speakers living in the different regions sampled. However, in many cases, these proportions are difficult to determine accurately. For example, it is difficult to determine exactly what proportion of the texts published every year belong to the fiction genre and how many are non-fiction. In this case, obtaining the exact figures is undoubtedly possible, but highly complex. The problem becomes even more challenging for the categories of unpublished texts, for which there are no existing figures. In these cases, the classification is often based on common sense or on the pragmatism of the corpus designer, depending on the importance of subcategories for addressing the questions that the corpus is supposed to help study.

Now, let us move on to the question of which samples to include in the corpus. The first important question is how these samples should be chosen. A first technique involves choosing the samples completely at random, the idea being that out of the total number of samples in the corpus, the most frequent characteristics will eventually stand out on their own. However, we cannot take for granted that this method yields a balanced sampling, especially in the case of a small corpus. As previously mentioned, a better method might be to define a sampling frame and to divide the samples to be collected depending on the important properties of such a frame. For example, if the sampling frame for a corpus of French spoken in Switzerland includes different criteria such as gender, age, socio-economic level or place of residence, an equivalent number of samples should be chosen to match each selected criterion. Within each criterion (e.g. 20- to 30-year-old middle-class men living in the canton of Geneva), participants can be chosen at random. According to Biber (1993), this technique, which is called stratified random sampling, never gives less representative results than purely random sampling, and most of the time, its results are much more representative than a random selection.

The next question to consider is related to the number of samples required in the corpus, as well as the ideal size for each sample. Once again, the answers to these questions depend on the type of corpus the researcher has in mind. The more generic the corpus, the more samples will be needed, whereas for a more specialized corpus, fewer samples are necessary in order to provide representative data. On the basis of corpus studies on the differences between genres and between language registers, Biber (1993) argues that in most cases, 10 samples of 20,000 words per genre are enough to obtain representative samples for each genre. Indeed, from this size onwards, the new occurrences found by incorporating additional samples become smaller in number. In the case of a reference corpus like the *British National Corpus*, 40,000 word samples were retained (Nelson 2010, p. 59).

Finally, another important question concerns sampling units. Should we include whole texts or only excerpts, or even isolated sentences? The answer to this question often depends on the accessibility of data. On the one hand, it is preferable to create a corpus including language samples which represent a coherent whole, rather than isolated sentences. However, this is not always possible due to copyright reasons. The correct size of samples also depends on the type of text considered. For example, if the goal is to reach a sample of 200,000 words per text genre, this number of words can almost be instantly reached by including one or two entire books in the sample. In this case, it would be a better idea to choose excerpts (e.g. chapters) from different books, instead of a longer portion of a single book. For other types of text such as letters, text messages, etc., the units are so small that it makes no sense to not fully include them. In all cases, it is important to systematically avoid including the same portions of text, for example, always the beginnings or the endings. Indeed, Stubbs (1996) observed that there are very few linguistic features which remain constant in a text. In order to observe all linguistic phenomena, it is therefore necessary to modify the text portion (beginning, middle, ending) that is included in the corpus.

6.4. Preparing and coding corpus files

In order to include language samples in a corpus, first we have to obtain them. In the case of spoken corpora, data acquisition first requires them to be transcribed. This is a very complex process, and we will discuss it in detail in the next section. In the case of written corpora, the situation is not always simple, either. The most favorable scenario is clearly the one in which data are readily available in digital format, which is progressively becoming more frequent, especially when it comes to data gathered from the Internet. However, retrieving text from digital formats may have varying degrees of difficulty depending on the original format; for example, it can be very difficult to isolate the texts of different articles in a journal page formatted as a PDF document, or even impossible when the PDF file includes text images. So, even when we are working with data in a digital format, the task of converting the original format into a usable format, based on the text, is still necessary, as we will see below. In some contexts, however, written data are not available in digital format. In this case, we can either work with printed texts available on paper or with handwritten texts, such as student essays or private letters. Printed texts can be scanned and then processed thanks to optical character recognition (OCR) software, but these always require a manual check made by a human in order to provide a completely reliable result. There might be a high number of errors if the original print is of a poor quality. Finally, for handwritten data, there is no solution other than to manually type it on the computer. Data transcription also raises many questions related to the way in which some of their original features might be preserved. For example, student essays often contain spelling mistakes, which should be left untouched, since they can be very informative for many research questions. But in this case, a version without misspellings should also be included so that the words can be found by a concordancer. In Chapter 7, we will see that errors can be systematically annotated in a corpus, in the same way as syntactic or semantic information is provided.

No matter the way of acquiring data, an important point is to save the corpus files into a format which can then be used by a concordancer. As we saw in Chapter 5, most concordancers (like AntConc) only read files in text format, which can include texts tagged in XML format. Therefore, all newly created files for a corpus should be directly saved into text format. Files which have already been scanned are rarely saved in this format, since this format does not make it possible to include formatting marks, and this makes documents difficult to read. In the case of corpus studies, this is not a problem, however, since the files will not be read by humans, but processed by a concordancer. Files processed with OCR software or word processing tools are often saved in proprietary formats (such as Microsoft Word, for example, DOC or RTF). Files saved in these formats can be easily transformed into text files thanks to specific options such as the "save as" command found in word processors.

Web pages are available in HTML format. This format contains many formatting marks in the form of tags, which are interpreted for creating the various graphic effects which are necessary for a browser to display a web page. These later become visible when a file is opened with an editor in pure text format. Despite their lack of linguistic relevance, these tags are interpreted as textual elements by concordancers.

In order to avoid this problem, software should be used for retrieving text from these files and eliminating unnecessary markings (HTML tags). Word processing software often include this feature, simply by choosing the "text format" option from the "save as" command. The only problem with this option is that it is necessary to open every file one after the other in the word processor so as to perform the operation. This might eventually become a problem with a corpus, including thousands of different files. An alternative solution is to use the AntFile Converter, which is a file conversion software developed by Laurence Antony, the creator of AntConc (see the URL at the end of this book for AntFile Converter). This software can be downloaded for free and used for converting any number of XML, HTML or sometimes even PDF files into text format.

The *Sketch Engine* corpus creation and management platform, discussed in Chapter 5, also automatically transforms the format of files downloaded from the Web. The advantage of this platform is that it offers the possibility of automatically downloading large amounts of data from the Internet in a single operation (web crawling). The corpus created in this way can be directly analyzed using the tools provided by the platform, for example, for retrieving concordances, word lists or keywords. In its recent versions, the WordSmith concordancer also offers a similar function. This type of tool has made the collection of web-based corpora extremely easy. We should nonetheless bear in mind that the texts found on the Internet are of a highly variable quality and are not representative of the whole language.

If the corpus has been created manually rather than through the use of a platform enabling automatic data download, a practical but important question concerns the number of files that should be created. More specifically, should we create a single file for the whole corpus? Or should we create one file per corpus sub-section? What about a file per text included in the corpus? In general, it is preferable to store every language sample in a separate file. In this way, it is later easier to combine them in different ways for creating sub-corpora, rather than having to retrieve text portions from a larger file. For example, if we collect data on the language used by young people in France, we might then want to compare data depending on different criteria such as gender, geographic region or age group. This comparison can be done in a relatively simple way by grouping all the men's files and all the women's files or, for the same purpose, all the Paris files and all of the Marseille files. But if all men are included in a single file and women in another, then the geographic comparison data needs to be reprocessed.

In order to be able to easily group the files into sub-corpora, it is necessary to represent the features of each sample in an easily accessible manner. A practical solution is to code these characteristics directly onto the file names: this is why these names are another important element that should be taken into account when creating the corpus. For example, it is possible to name files only by using a number, each representing a criterion used when compiling the corpus. Going back to the example of young speakers, one possibility would be to identify all the files from Paris with the number "1", those from Marseille with the number "2", etc. Then, the second digit could be used for coding gender, "1" for women and "2" for men, then the third reference could be for coding the age group, for example, "1" for 16- to 19-year-olds, "2" for 19- to 22-year-olds, etc. Finally, several digits can be used for coding the participant's number. Following this procedure, the sample corresponding to the first 18-year-old male participant from Marseille registered in the corpus would be saved in a file called "221001.txt". The disadvantage of this method is that the coding is opaque for a user who does not have a precise vision of the system used.

A more transparent way to achieve the same result is to use abbreviations. For example, the same file could be coded using abbreviations such as *Mar* for Marseille, *h* for men (*homme*), *ado* for the 16- to 19-year-old group, which would result in a file called "mar_h_ado_001.txt", if we use the underscore symbol as a separator for the abbreviations. If this system is used, abbreviations should be kept short in order to avoid generating excessively long file names, which might not be readable. We should also avoid inserting spaces or other punctuation marks, since these could interfere with the programs used for opening the files on different platforms (typically concordancers). Finally, if word abbreviations are used, it is desirable that each abbreviation of a category contains the same number of characters (e.g. three letters for all the names of cities), in order to make reading in columns of lists of files easier.

We have pointed out that corpus files should contain plain text, in order to facilitate data analysis. However, for a corpus file to be used as a sample representing a certain type of language, metalinguistic information (which is not part of the text or of the dialogue) should be accessible to the researchers who will analyze it. For example, this type of information includes the date of a newspaper article, the place where a conversation was recorded or the characteristics of the speakers taking part in the dialogue. This "piece of extra information concerning the data" included in the corpus is what we call metadata.

For this information to be made available for future users of the corpus without separating it from the rest of each sample portion, it should be possible to include it in the files, but in such a way that it is not taken into account by a concordancer when counting the words of the corpus. A possible solution could be to insert these marking elements inside tags, something that the concordancer will be able to ignore. Most often, these tags are delimited by chevrons (the less-than and greater-than signs < >).

In this way, the metadata of a corpus sample can be added at the beginning of each document as follows:

<texttype: newspaper article>
<publication: *Le Monde*>
<author: Jean Dupont>
<date: 1 April, 2019>
<subject: April Fool's Day>

In the AntConc concordancer, discussed in Chapter 5, it is possible to inform the program about the existence of tags and not to consider the information they contain.

In some cases, the abundance of metadata requires the use of a precise syntax for tags, based on the conventions of computer languages for XML or SGML coding. The conventions used can be explained in the corpus documentation or may follow a more widely recognized standard. Indeed, some sophisticated marking formats have already been developed for corpus data. One of the best known is the TEI format (Text Encoding Initiative), which makes it possible to encode many markings in a standardized way and make corpora sharing easier. Large reference corpora such as the *British National Corpus* are tagged following the TEI conventions. Without going into details, a TEI-tagged document always contains two types of elements:

– the header;

– the body of the text.

And these two elements are respectively made up of other elements. The header section contains metadata, a description of the file, the encoding, the text profile, mainly the language, the context or participants, and even a history of its revisions. All these elements, except for the file's description, are optional. The body of the text mainly contains tags, which are destined to delimit text units such as paragraphs or even sentences. Following XML conventions, TEI tags always begin with chevrons < > and close with </ >. As we will see in Chapter 7, TEI tagging can also be used for making more detailed annotations than dividing the text into sentences.

In addition to indicating the metadata by means of tags inside each file, it is also very useful to provide a summary table in the corpus documentation, as illustrated in the simplified Table 6.2. This type of table gives users an idea of the contents of the corpus at a simple glance and helps them to quickly choose those files which are relevant to their concerns, without having to open them all one by one.

File	Gender	Age	Residence	Context	Topic
113001.txt	Male	28	Neuchâtel	At home	Retells a memory
224002.txt	Female	32	Geneva	At work	Talks about her work
212003.txt	Female	25	Martigny	At a coffee shop	Retells a memory

Table 6.2. *Example of a table summarizing corpora metadata*

6.5. Recording and transcribing spoken data

The collection of spoken corpora poses certain additional challenges compared to written corpora. One of the main difficulties stems from the need to transform spoken data into a written format. As we have seen in Chapter 5, for corpus data to be analyzed, they should be in written format, since concordancers cannot search for words or expressions in audio files. In this section, we will briefly discuss some of the problems related to the representation of spoken data, as well as some possible solutions to sort them out.

The first step we can take to work with spoken corpora concerns the mode of acquiring data. Spoken data need to be recorded and the recording process itself requires special preparation. For data to be as representative and informative as possible, it is essential to properly define the research questions that these data will help answer well in advance. In particular, these questions will determine the type of interactions that should be recorded, the constraints regarding the context as well as the information contained in the transcript. If, for example, the aim of a spoken corpus is to study the lexical specificities of a language variety, the prosodic information contained in the interactions will be of little use. If, on the other hand, the research question concerns information structure in discourse, more specifically the introduction of new and given information in different spoken genres, then prosody will play an important role in studying the interface with the utterance structure and therefore requires a transcription.

An important point to establish before carrying out the recordings is the nature and the amount of contextual information that will need to be added to the transcripts. Spoken conversations are naturally more ambiguous and less precise than written communication, since speakers can use the immediate context to make themselves understood. Audio recordings make it almost impossible to grasp this type of information, which later have to be added to the corpus so that the interactions can be understood and analyzed by the experts who will listen to these recordings. In the case of audiovisual recordings, a larger share of contextual information will be captured and should not be explicitly added to the corpus (although some kind of codification may later be required to perform specific analyses).

Due to the difficulty of collecting and transcribing spoken data, the question of the amount of data needed for creating a corpus is even more acute than for written data. But in this case too, there is no ideal size for a spoken corpus. The amount of data required for studying a certain phenomenon primarily depends on how frequently it occurs. For example, Adolphs and Knight (2010, p. 41) have estimated that one hour of recorded conversation corresponds to approximately 10,000 words in the transcript.

As we have already mentioned, providing metadata details is particularly important in the case of spoken corpora. Among other things, the metadata appearing in the header should offer information about the main features of the participants: degree of relatedness (parents, friends, colleagues, strangers, etc.), the context in which the conversation took place, the manner in which the recording was captured, etc. The importance of this information depends on the questions that the corpus is expected to answer. For studies on how interpersonal relationships influence interactions, it will be necessary to have as much information as possible about the degree of relatedness between participants, whereas for studies on the use of discourse markers such as *bon* or *ben*, this type of information is not so relevant.

In the same way, contextual information may be added to the statements inside the transcripts. Let us insist on the fact that the context of an interaction is so rich that it would be illusory to try to account for all the aspects involved in a transcript. Choices will have to be made depending on the importance of this information for the research question. At least, the transcripts should contain enough contextual information for the meaning of the utterances to be reconstructed if this became necessary in the absence of context. For example, if a person passes by and this event invites a comment from the participants, this piece of information should be mentioned in the transcript, indicated between tags so as not to be confused with the transcription itself.

Finally, the last difficulty related to transcription that we will mention concerns the presentation of the transcripts. In a dialogue, the participants do not always speak one after the other as it happens in the dialogues of a novel or a play. On the contrary, there are many overlaps between speaking turns, as well as pauses. The analysis of overlaps and pauses can be important for certain studies, so the question arises on how to best account for these phenomena. If a transcript is presented in a purely linear fashion, one intervention above the other, valuable information might be lost. This is why other types of presentation are often used. For example, in the CLAPI corpus, the contributions from the different participants are presented one below the other. Overlaps are indicated by green square brackets, making it easy to see where and when they occur, as shown in Figure 6.1. Numbers in light blue indicate the presence of pauses and their duration in seconds.

```
JEA bah c'est (inaud.) tout façon
JUL c'est mignon ouais/
JEA [ah oui]
JUL [nan mais][la (inaud.) c'est quand même différent en france tu vois]
LAU            [non mais linköping lin- linköping c'est                 ]
    plus p`tit que besançon j` veux dire
(1.1)
JEA j` veux dire b`san::çon
JUL ((rire))
CLA < ((en riant)) huhuhuhum
```

Figure 6.1. *Example of a CLAPI corpus transcription. For a color version of this figure, see www.iste.co.uk/zufferery/corpus.zip*

Yet another solution is to represent the words from each participant in a separate column and to show the overlaps on the same line.

In summary, the transcription of spoken data requires many decisions to be made concerning the nature and the amount of information to be added, not only to the dialogues themselves, but also on how to communicate such information on the files and visually. These decisions should be made even before the data collection process begins, since an important portion of contextual information could be lost if it is not recorded during the interactions.

6.6. Ethical and legal issues

Creating a corpus involves using (or even sharing with other researchers) language samples produced by third parties. Those persons having contributed to a corpus through their language productions have rights that need to be respected. In the case of a spoken corpus in particular, it is essential for participants to know that they are being recorded and that their data will later be used for linguistic analyses. For this, the creators of a corpus must hand a document to their future participants clearly explaining who the data will be accessible to and how it will be used. The participants can then freely decide whether or not to sign a form stating their consent to take part or not in the study. However, such consent to participate does not suffice to share the data with other people afterwards unless this usage has been explicitly mentioned in the form. In fact, a participant may agree with the idea of being recorded by a researcher and then having such data used for research, but not necessarily agree with having their data being shared with a large number of people, perhaps even with web-free access. In order to be able to share corpus data, it is imperative to ask participants both for their authorization to use and to distribute the data before collecting them. If a participant later refuses to have their data distributed, removing them from the corpus may pose many difficulties. In the case of dialogues, all of the data of the events that the participant was involved in will have to be removed.

The right to anonymity of the persons mentioned in the corpus represents another important ethical problem. Often, people interacting in recorded conversations refer to third parties by naming them. These people did not provide their consent to being talked about in public documents, so their names should be removed before publishing such data. This anonymization process is not always easy, however. For example, McEnery and Hardie (2012) have mentioned several cases drawn from the *British National Corpus* in which the people in question were very easily identifiable even after their names had been deleted, since the context was precise enough to be able to find them on the Internet. For instance, this is the case of conversations in which persons with a specific role in the village, such as the doctor or the clergyman, are mentioned. This situation is particularly problematic when references to people include degrading criticism or reveal their illicit activities. In these situations, it is necessary to delete parts of the conversation, or even the entire conversation, to protect the rights of the persons concerned.

In the case of written corpora, the situation is simpler, especially when it comes to published data. It is reasonable to think that the public figures mentioned in the articles agree to waive their right to anonymity. The responsibility of the corpus compiler is involved when it comes to texts with potentially defamatory content. In the case of articles found on the Internet, in particular, source verification is necessary before indiscriminately including texts collected automatically, following the web crawling processes described earlier in this chapter. In general, we should also be aware of the fact that distributing a corpus implicitly amounts to disseminating the ideas contained inside its texts. In some cases, this may pose ethical problems for researchers. For example, Baker and Vessey

(2018) have compiled a corpus of propaganda by extremist groups. Distributing such a corpus clearly invites the question of whether the ideas it contains should be propagated on the Internet or not. The authors were also unable to access a portion of the propaganda journals published by certain terrorist organizations. In the United Kingdom, the possession of this type of material is illegal. This example brings us to the second aspect considered in this section, namely the legal issues involved in the creation and distribution of corpora.

Written corpora containing published texts are confronted with copyright issues. While it is true that laws differ from country to country, it is not possible to distribute the texts of an author during his/her lifetime without demanding some type of compensation. In France, this period is valid until 70 years after the author's death. However, contrary to what many people think, data accessible on the Internet are also subject to copyright. Their use can be softened, providing that they are accompanied by sufficiently permissive user licenses, such as the Creative Common license which concerns the contents of the collaborative encyclopedia Wikipedia. Due to copyright restrictions, in the case of less permissive licenses, corpora creators encounter many restrictions for including data. There are several possible strategies for properly addressing the copyright problem.

First, we can limit our choice to works that have fallen into the public domain and/or coming from websites where data have been declared free of rights. This solution is the safest one from a legal point of view, though it is not the most satisfactory one from a linguistic point of view. As a matter of fact, this selection method hinders the collection of data that truly mirror contemporary language and certain stylistic genres which are poorly represented on the Internet.

A second solution would be to negotiate the right to use data with their owners. This solution can be realistic when creating a corpus drawn from a limited number of sources. Rights holders often agree to authorize a single researcher to use a reasonable amount of their data for research, but this type of corpus often cannot be later redistributed. This limitation poses a problem for research replicability, which is an important scientific element in order to grant its validity.

Another way of dealing with the copyright problem during corpus distribution would be to allow users to search for concordances in the corpus, but not to visualize it in its entirety. This is the case for many corpora that are only available via an online consultation interface. These interfaces only enable occurrence searches for words or expressions within a certain context. This solution effectively preserves copyright, since the works remain inaccessible. For users, these interfaces make it possible to answer a certain number of research questions related to the lexicon.

However, they are unsuitable for research questions that require data processing, for example, some kind of annotation or those that have to take into account a large context, in order to identify certain linguistic phenomena such as speech acts or discursive phenomena.

6.7. Conclusion

In this chapter, we have discussed the main elements to consider when creating a new corpus. For a start, we mentioned that corpus creation is a long and complicated process. This is why reusing the already existing data should be prioritized as far as possible. Then, we saw that the important methodological trait to be respected when creating a corpus is data-representativeness. The latter can only be defined in relation to a specific research question. The representativeness of a corpus also depends on its balance and the choice of samples it contains. We also introduced some basic principles regarding sample collection and balancing. We then addressed some concrete problems, related to data coding and transcription into a corpus, and concluded that these questions needed to be resolved before starting the data collection phase. Finally, we saw that the creation of a corpus poses several ethical and legal questions which should be carefully considered, since distributing data amounts to disseminating information that belongs to and concerns third parties, whose rights must be respected.

6.8. Revision questions and answer key

6.8.1. *Questions*

1) What types of data should be collected to conduct a representative study of how young people use the discourse marker *genre* in French?

2) How could we balance the different parts of a corpus aiming to study the French literature of the 18th Century?

3) What are the main questions to consider when choosing the samples to be included in a corpus?

4) Using the *Sketch Engine*, choose five keywords in order to create a corpus on the *French cinema*. What are the characteristics of the corpus thus created? Which are these keywords?

5) What transcription information would it be important to add to a corpus of spoken conversations to study the language of the suburbs in France?

6) What are the ethical issues to consider in the following cases:

a) a collection of texts produced in class by children;

b) a recording of spontaneous conversations of a group of friends at a bar;

c) a recording of a teacher's course for a spoken corpus.

7) Which of the following actions do you find problematic from a copyright perspective:

a) using a digital version of the novel series *Harry Potter* for compiling a corpus stored exclusively on your computer;

b) sharing this corpus with your partners as part of a corpus linguistics course in order to do joint homework;

c) distributing this corpus on the Internet;

d) including an entire chapter drawn from this corpus in a publication with the aim of illustrating certain linguistic phenomena that you have annotated.

6.8.2. *Answer key*

1) In order to have representative data for this research question, the corpus chosen should evidently contain language produced by young speakers. This concept would need to be clarified to be operational, for example, by deciding to include an age group ranging from 15 to 25 years.

This study does not specify a geographic region. One way of delimiting research for such a project would be to compare young people living in large cities in four French-speaking regions from different countries, for example, Paris, Brussels, Geneva and Quebec. The young speakers included in the corpus should proportionally represent the two genders and have different socio-economic profiles. Finally, the corpus should contain young speakers recorded under similar conditions in order to avoid any context-related bias. Another study could aim to compare these uses across different speech styles, which should then be represented in the corpus in a balanced manner.

2) This research question is fraught with different constraints. First, the corpus should contain French literature, which restricts the literary subject to works written in original French, rather than translations. It should also span a specific period, which could be defined, for example, as works published between 1800 and 1900. The difficult point to assemble this corpus relates to the way of balancing its content between different literary genres. It should therefore contain novels, short stories, plays and poetry. Let us assume that the corpus targets a size of 200,000 words per genre, in order to have a representative sample of each of them. Since novels and plays are long texts, it would be a good idea to include excerpts (e.g. a chapter or an act) from many different works, rather than two or three texts in their entirety. Conversely, since poems are a very short genre and more marginal in terms of the amount of texts published, a possible decision would be to limit the share of poems to a smaller percentage of the corpus, for example, to limit poetry to 50,000 words.

3) The first question to ask is whether a sample is representative of the genre it embodies in the corpus. For example, before deciding to include an interview with a young Brussels resident in a corpus of French spoken in Belgium, we should make sure that this person has not recently moved to Belgium from another country, and that they properly reflect the linguistic specificities that the corpus is supposed to embody. The second important question concerns the size of the sample that will be included in the corpus. As we recalled above, it is not always optimal to include entire texts in a corpus when these are very long. Depending on the target size for each genre, it is necessary to determine the appropriate size for each sample. A third question to consider concerns the way in which the samples are acquired, depending on whether these are digitized texts, texts to be scanned or transcribed. Besides, it is also necessary to determine which metadata will

be associated with each corpus sample. All of these decisions need to be made based on the research question being considered. A final and very important point is to ensure that copyright is respected, either because the text is copyright-free or because the author has provided written consent granting permission to include their text in the corpus. Finally, we should make sure that the sample is acceptable from an ethical point of view, in the sense that it respects the right to anonymity of the person involved and that it does not contain inappropriate content.

4) By creating a corpus with keywords such as *cinéma*, *films* and *acteurs* and using the default parameters offered on the site, *Sketch Engine* produces a corpus of 90,441 word occurrences, including 2,680 word types retrieved from 35 different pages. The most frequent content word is *cinéma*, at the 20th frequency rank. The keywords in the corpus include proper nouns such as *Edison*, *Funès*, *Fernandel*, *Gabin* and *Reynaud* and also content words like *cinéphile*, *cinéma* and *crédits*. Collocations include elements like *cinéma français*, *cinéma muet*, *art dramatique*, *histoire du cinéma*, *grand écran*, *carrière cinématographique*, *mise en scène*, *film français*, *actrice américaine*, etc. These collocations make perfect sense in view of the search terms used for creating the corpus.

5) Transcription information should include both metadata and indications inside the transcripts themselves. The metadata of such a corpus should at least include information of where the recording took place, its date, the context in which it happened, the conversation topic, the number of participants, the gender of the person recorded, his/her age and profession. Within the transcripts, it might be useful to include an annotation of the non-standard words used, for example, in verlan, with their equivalent standard so that they can be found in a concordancer. An indication of pauses, overlaps and certain prosodic phenomena may also be useful.

6) a) Above all, a **collection of texts produced in class by children for assembling a corpus** requires protecting the children's right to anonymity. No element in the corpus should make it possible to identify any participant. Depending on the nature of the texts, it is necessary for the content to exclude any element making it possible to identify any other person.

b) In the case of a **recording of spontaneous conversations of a group of friends at the bar**, everyone involved should be notified that the conversation is being recorded and that he/she agrees. Depending on the purpose of the corpus, this agreement should also consider data sharing with third parties. Then, depending on the conversation topics, it would be necessary to ensure that the content is neither defamatory nor offensive, and that it does not enable third party identification.

c) Finally, in case we decide to **record a lesson from one of the professors for a spoken corpus**, we should make sure that the professor has been informed about the recording and has given his/her consent, both for the use and for the possible sharing of the data.

7) a) Using a digital version of the *Harry Potter* series for assembling a corpus to be stored only on your computer and searching for elements in the text is not problematic *a priori*, insofar as this digital version has been legally acquired, for example, by buying an e-book from an online bookstore.

b) However, sharing this corpus with your classmates within the framework of a corpus linguistics course with the aim of carrying out joint homework is a bit more delicate an issue, because the fact of buying a book does not entitle you to duplicate it or to transmit it free of charge to others. This practice may, however, be considered as a tolerated use of the material if the aim is to carry out joint homework on the data, which are not being used in any other way.

c) Distributing this corpus on the Internet is completely illegal under copyright rules which can be enforced for decades after the author's death (70 years in France). In this case, the *Harry Potter* series will still be protected for many years, and any form of distribution is currently prohibited.

d) Including an entire chapter of this corpus within a publication to illustrate certain linguistic phenomena that you have annotated can also be problematic from the point of view of copyright. Though it is acceptable to quote portions of a text while indicating its source, these quotations should not exceed a clearly defined size limit. This size varies from country to country, but generally does not exceed a few hundred words. Thus, publishing an entire chapter of a book is not acceptable.

6.9. Further reading

Wynne (2005) addresses the different stages and difficulties associated with corpus data collection in an accessible but detailed manner. The methodological principles related to the creation of a corpus are discussed in detail by McEnery *et al.* (2005, section A) and more succinctly by Nelson (2010). Koester (2010) includes practical advice for the creation of a small specialized written corpus and Adolphs and Knight (2010) for the compilation of spoken corpora. The ethical and legal questions associated with the creation of a corpus are addressed by McEnery and Hardie (2012, Chapter 3).

7

How to Annotate a Corpus

In Chapter 6, we discussed the importance of associating metadata with corpus files. In this chapter, we will explore how to insert other types of information into a corpus as linguistic annotations. To begin with, we will see that annotating a corpus is a way of adding value to it by widening the field of questions that it will make it possible to investigate. We will then review the different types of annotations we can add to a corpus, briefly present some tools for performing some annotations automatically or for making manual annotations easier. We will also discuss best practices to follow when making a manual annotation and present the different ways to assess the reliability of such annotations. Finally, we will present the principles to be respected in order to make annotation sharing easier.

7.1. Corpus annotations

Raw data which are collected in a corpus are not always adequate for answering many research questions. Let us imagine, for example, that we wish to know how often French nouns such as *ferme* and *car* appear in a corpus. A simple search for their occurrences in a raw data corpus will provide a biased answer, since these words also have other uses apart from being nouns, and these will be included among the search results (*ferme* is also a conjugated form of the verb *fermer* and *car* is also a coordinating conjunction). In order to keep only the relevant occurrences, should we filter all the relevant occurrences by reviewing them one by one? Such manual sorting tasks can actually be avoided if the words in the corpus have been previously annotated into grammatical categories. This annotation makes it possible to exclusively look for the noun occurrences of *ferme,* for example.

The great advantage of part-of-speech tagging is that it can be done automatically with almost the accuracy of a manual annotation, regardless of the amount of text to be annotated. As a matter of fact, part-of-speech tagging has been performed on the *Google Books* corpus (see Chapter 5), which contains billions of words from different languages and has made it possible to refine research on language evolution. For example, Lin *et al.* (2012) found that the regular past participle of the verb *burned* replaced the irregular form *burnt* several decades before the change affected its adjectival uses, thus suggesting that verbs played a special role in this evolution. However, such a role could not have been detected by looking for word occurrences in a non-annotated corpus.

Despite these advantages, annotating corpora also has its drawbacks, which has made some researchers prefer to stick to the analysis of raw corpora (see, for example, Sinclair 1992; Hunston 2002). These researchers have particularly criticized annotations because they make the corpus less readable and less flexible. It is true that if, in a syntactically tagged corpus, every word can only be viewed followed by a code showing its grammatical category, such as "mange_V3PP" to indicate that it is a verb in the third person of the present, the corpus becomes difficult to read. To avoid this problem, it is essential for annotations to remain separated from the corpus, which makes the corpus always accessible as plain text. This problem is particularly acute in the case of monitoring corpora for which it becomes more complex to regularly add new texts, since these must follow the same annotations as previous ones. For these corpora, automatic annotations as part-of-speech tagging, for example, are a good option and offer added value to their content.

Finally, the main objection often raised against annotations is that any form of annotation is necessarily subjective, at least in part. This implies making choices about the categories to be annotated (see section 7.3), which are never completely neutral. Furthermore, the annotation itself is accompanied by a hermeneutic dimension, which reflects the annotators' point of view about the text. This is due to the fact that all the categories include borderline cases, for which different annotations can be justified. In these cases, annotators must make decisions while simultaneously interpreting data.

An example drawn from recent research on language acquisition will better illustrate the impact of annotation choices. Different authors have studied the early production of causal connectives in children, in different languages. Kyratzis *et al.* (1990) decided to retain and to annotate all of the children's utterances containing a causal connective, even when children produced only part of the sequence themselves, for example, when answering a question, as Léa did at the age of 2 years old in (1), in the *York* corpus (retrieved from the CHILDES database, see Chapter 5).

(1) *Grand-mère: pourquoi ça?* (Grandmother: why that?)

Léa: parce que j'en avais envie. (Léa: because I felt like it.)

Conversely, Evers-Vermeul and Sanders (2011), as well as Zufferey (2010), decided to keep only the occurrences in which the entire causal relation had been produced by the child. This methodological difference has generated different conclusions concerning the order of acquisition between the different types of causal relations. Kyratzis *et al.* (1990) found that children first produced subjective causal relations (see Chapter 2, section 2.5), since these often occur in question–answer sequences. In striking contrast, Evers-Vermeul and Sanders (2011), as well as Zufferey (2010), found that children first produce objective causal relations for describing facts or events in the world as in (2), produced by Léa at 2 years and 10 months old. In fact, children do not produce subjective causal relations entirely by themselves until later, in sequences like (3), produced at 3 years and 10 months old by Léa, in the *York* corpus.

(2) *Léa: je prends ma gourde parce que j'ai soif.* (Léa: I'm taking my bottle because I'm thirsty.)

(3) *Léa: venez parce qu'il est très tard !* (Léa: come because it is getting very late!)

This example illustrates the influence of the methodological choices associated with data annotation and the conclusions that can be drawn from a corpus study.

While it is true that annotation processes involve choices that are always partly subjective, many researchers (e.g. McEnery *et al.* 2005) have argued that any corpus analysis, even when performed on raw data, implies such choices and no analysis can therefore be considered as completely objective and neutral. Furthermore, annotations have the merit of presenting these choices in a transparent manner when the annotated data are shared (see section 7.6). They also enable other researchers to comment on them, or even revise them. Going back to the example of causal relations, the differences between studies do not blur the fact that children begin by producing partial causal relations as a reply to questions, later produce objective relations by themselves and finally, subjective relations. These occurrences are discussed in all studies, and later integrated or eliminated from the data to be annotated. Finally, raw data cannot simply be used for handling many research questions (see Chapters 2–4) without the use of annotations, since this would imply a sharp limitation to the field of corpus linguistics. Conversely, the same annotation can often be useful for approaching several research questions (see section 7.2) and the effort invested during the annotation process may add lasting value to the corpus. This is why the majority of researchers are in favor of corpus annotation, provided that the task is carried out while respecting some principles, which we will detail in this chapter. But before that, let us see how annotations can adopt very different forms depending on the area of language study.

7.2. Different types of annotations

All levels of linguistic analysis can be annotated in a corpus. To begin with, some annotations are related to phonology, particularly in the case of transcriptions for spoken corpora. For example, these annotations indicate prosodic phenomena like pauses, hesitations and prosodic phrasing. In this way, they make it possible to look for such phenomena automatically, without having to listen to all the audio files in the corpus again. In addition to the study of prosody, these annotations can be useful for works of a very diverse nature, for example, for studying the notion of fluency or the interface between syntax and discourse, so as to better understand information structure in discourse.

Words are often the linguistic element the most subjected to annotations in a corpus. The most basic of these annotations is the division into words or occurrences (known as *tokenization*), which includes punctuation identification, elision processing (in French) or the identification of numbers and dates. However, lemmatization refers to the act of associating every word occurrence in a corpus with its basic morphological form. For example, adjectives such as *gentil* or *gentille* (in French) are the masculine and feminine variants of the same word (or type) from a morphological point of view. This canonical form of the word is called its lemma. In the case of adjectives, their lemma is by convention the singular masculine form. Similarly, the words *mouse* and *mice* are the singular and plural variants of the same lemma (*mouse*), whereas *eat*, *eaten* and *eating* are the conjugated forms of the lemma *eat*. Lemmatizing a corpus is an essential process for studying lexicon, since annotation makes it possible to look for occurrences without having to detail all the morphological variants it may adopt. Lemmatization is one of the types of annotations which can be done automatically, with considerable accuracy (see section 7.4).

The notion of the lemma should be differentiated from the concept of the lexeme, which is used for defining a word from a semantic point of view. So, for example, *bat* is a lemma which can correspond to two different lexemes, as this word is polysemic and may either refer to a flying mamal or an object used to hit a ball. Finally, we should bear in mind that the usefulness of lemmatization may change significantly from language to language. In languages making little use of inflectional morphology such as English, this process is not much useful. It is nonetheless more useful in French, and even more in languages containing an abundance of morphological inflections, such as Finnish or Russian. In addition to lemmatization, words can be annotated into grammatical categories thanks to part-of-speech tagging, as we previously mentioned in relation to word annotations such as *ferme* and *car*. We will come back to this later in the chapter.

Finally, words can be annotated into semantic categories, for example, the word *tennis* can be annotated as a part of the *sports* category, later making it easier to quickly go through the content of a corpus, and to disambiguate polysemic words in context. Annotation also provides training and testing data for automatic word sense disambiguation. Indeed, this type of annotation is more difficult to perform automatically than part-of-speech tagging, since it requires conceptual knowledge in context and this is still a major challenge for artificial intelligence. For example, depending on the context, the word *mouse* may belong to the category of computers or animals.

In addition to words, sentences are also regularly annotated in corpora. The most common annotation is syntactic parsing. As soon as a corpus has undergone a part-of-speech tagging process, it is possible to parse it, and thus reveal how grammatical categories can be grouped into smaller phrases within a sentence. In traditional grammar, these syntactic representations often adopt the form of constituent analyses, as illustrated in Figure 7.1 for sentence (4).

(4) *The teacher congratulates the student*

Due to their tree representations, parsed corpora are often called "treebanks". The most famous corpus containing such an annotation is the *Penn Treebank* corpus (see Taylor *et al.* (2003) for an overview of this English-language corpus), produced at the University of Pennsylvania and including a syntactic annotation of 3 million words from a corpus of 7 million words from different genres. This type of annotation was also carried out in French by Anne Abeillé's team in Paris on approximately 1 million words from the *Le Monde* corpus (Abeillé *et al.* 2003).

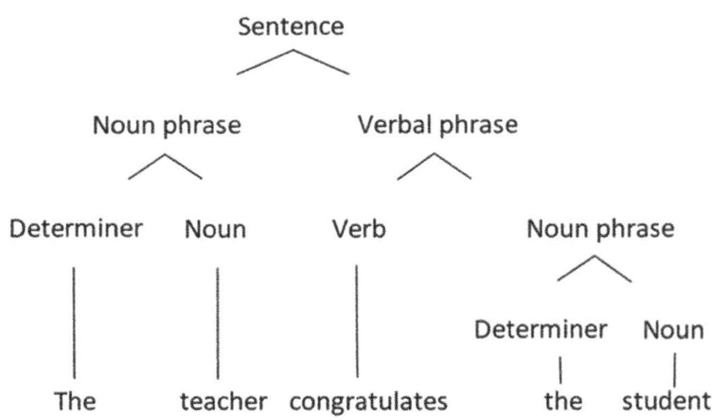

Figure 7.1. *Syntactic representation of a sentence in the form of a tree*

At this stage we should observe that syntax trees cannot be represented as such in the computer files containing annotations, but need to be codified in a way that is close to the original text. A frequent form of representation is based on the use of brackets or parentheses to delimit phrases. When these brackets are accompanied by tags (e.g. as super index after the closing bracket), the representation obtained contains the same information as a syntax tree, as illustrated in (5) for the tree in (4). Even if this representation is not particularly legible, the appropriate computer tools can make its graphical presentation more user-friendly.

(5) [[[The]$_{\text{Det}}$ [teacher]$_{\text{N}}$]$_{\text{SN}}$ [[congratulates]$_{\text{V}}$ [[the]$_{\text{Det}}$ [student]$_{\text{N}}$]$_{\text{SN}}$]$_{\text{SV}}$.]$_{\text{Ph}}$.

Apart from indicating constituent structures, some parsing analyses show the dependency relations between words or syntactic constituents. A dependency relation takes place when one word governs another. For example, in a noun phrase, the noun governs the adjective, which depends on it for receiving agreement marks. Dependency analysis was performed automatically on the *Google Books* corpus (Lin *et al.* 2012). This type of annotation makes it possible to explore the content of the corpus more in-depth than part-of-speech tagging, since it makes it possible to look for all the words experiencing some kind of dependency relation. In this way, for example, we can determine which adjectives are used the most often for modifying the word *hair*, such as *long* or *grey*, regardless of their exact position. Dependency analysis can be performed automatically, just as with constituency analysis, but its quality is lower than that of part-of-speech tagging. For the French language, Lin *et al.* (2012) have announced a 97.3% accuracy rate for part-of-speech tagging against only 84.7% for dependency analysis. We should point out that the system's accuracy rate may vary from language to language. At present, English has the highest accuracy rate, since research has mainly focused on this language. Conversely, many minority languages lack parsing tools.

Sentences can also be analyzed from the point of view of semantic relations between their constituents, as well as the thematic role of each of them. These roles include the agent, the patient and the cause. Finally, sentences may contain a pragmatic annotation of the speech act involved (e.g. a question, a request or a confirmation).

At a higher level, some annotations indicate the relations holding between sentences, or between discourse segments. In this case, it is rare to be able to rely on automatic annotation tools. For this reason, discursive annotations are most often the result of human annotators. These annotations include, for example, coreference relations, associating pronouns or noun phrases with their antecedent in discourse, as illustrated in (6). In this example, the pronoun *he* in the second sentence is coreferential with the noun phrase

Fred. In other words, it is used for designating the same person. This annotation is useful since a pronoun like *he* is not an autonomous referential expression, meaning that it does not by itself make it possible to identify the referent in question if we ignore the context.

(6) *Fred was very happy.* ***He*** *had won his tennis match.*

Other discursive annotations can explain discourse relations holding between phrases or propositions within complex sentences, such as causality as in (7), but also condition, contrast or temporal succession.

(7) *Fred was very happy* ***because*** *he had won his tennis match.*

In French, the *DEDE* corpus (Gardent and Manuélian 2005), available on the Ortolang platform (see Chapter 5), contains an annotation of anaphoras, whereas *ANNODIS* is a discourse-level annotated corpus (Afantenos *et al.* 2012). Finally, *Rhapsodie* is a 33,000-word corpus of spoken French, which is available online and contains syntactic and prosodic annotations (Lacheret-Dujour *et al.* 2019).

Finally, corpora including children's language or productions of foreign language learners may contain an annotation of errors. These annotations provide valuable clues to assess the nature and extent of the problems associated with each acquisition stage. Error annotations often include categories such as: incorrect word or structure, missing element, superfluous element or inflection/derivation mistake (see Lüdeling and Hirschmann 2015).

7.3. Standardization of annotation schemes

For all the annotations described in this section, international standards have emerged in the literature. When starting an annotation project, it is advisable to gather information about the existence of such standards and to consult previous studies to decide on the best way to annotate data. Whenever possible, it is preferable to reuse existing annotation schemes and to adapt them to the needs of the study rather than creating new schemes from scratch, for which comparisons with literature could be difficult to draw. One of the main initiatives, within the formal framework of the International Organization for Standardization (ISO), more specifically in technical committee no. 37 dedicated to language resource management (see the URL at the end of this book), has enabled around 20 standard drafts for linguistic annotation, be it lexical, syntactic, semantic or discursive in nature. For example, Bunt *et al.* (2012) developed an ISO standard (24617-2: 2012) for dialogue act annotation, a concept which includes both speech acts in a pragmatic sense and other conversational interactive elements, such as hesitations.

In many cases, standards are established *de facto*, when a big project is successful, rather than through the creation of a formal framework. This is the case, for example, in the field of discourse relations. The taxonomy of relations developed for annotating the *Penn Discourse Treebank* (Prasad *et al.* 2008), a version of the *Penn Treebank* to which discourse relation annotations have been added, is now widely used. The taxonomy used for the ANNODIS project in French has been inspired by another theoretical model of discourse relations, the *Segmented Discourse Representation Theory* (Asher and Lascarides 2003), which is also widespread.

Finally, we should point out that the annotations described in this section are performed on entire corpora. Indeed, each word belongs to a grammatical category and every sentence communicates a speech act. These annotations imply a global processing of the corpus. Conversely, many other annotations made on corpus studies are devoted to a single type of element to be studied. For example, Crible (2018) annotated the use of discourse markers in French and in English comparable corpora, without carrying out a complete discursive annotation. Likewise, the studies quoted at the beginning of the chapter have exclusively annotated the causal relations produced by children. These annotations were made depending on predetermined criteria for each study, rather than depending on formal standards. What is more, these were often not integrated into the corpus itself, but carried out in a separate file after retrieving the relevant elements from the corpus (see section 7.3). Most of the time, these were not as widely shared and reused as the original corpus. However, these somehow narrower annotations have become essential for many research questions, as we illustrated in Chapters 2–4.

7.4. The stages of the annotation process

In this section, we will detail the different steps which outline the process for annotating corpus data. Before starting the annotation process itself, we first have to define the categories which will be annotated in the corpus. In practice, these will be represented by a list of tags or sometimes, by dependency relations. For example, the annotation of speech acts requires setting up a list of the acts to be annotated, for example, requesting, promising, asserting, threatening, etc. However, a semantic annotation of verb types could differentiate their aspect (state or event verbs). Finally, an annotation of the grammatical categories associated with each word requires establishing a list of these categories. Defining categories is not as simple a process as it may seem at first glance.

One of the problems is the need to find a good balance between tags which are specific enough so as to avoid unwanted amalgams, and categories which are sufficiently generic so that they can be reliably identified by annotators (see section 7.5). For example, should we make a generic category including all the determiners or should we use specific tags for articles (*the*, *a*, etc.), on the one hand, and the possessive pronouns (*my*, *his*, etc.), on the other hand? Should we make a generic category for conjunctions or a category for coordinating conjunctions and another one for subordinating conjunctions? There is no single answer to these questions, as the right level of granularity of a (more or less precise or generic) annotation depends, to a large extent, on the objectives of the annotation and its feasibility. It is possible that some sub-distinctions are not useful and can therefore be omitted for a certain research question. For example, a corpus that is annotated into grammatical categories with the aim of being used as a reference tool for beginner language learners will have to use a simplified categorization of grammatical categories. Contrary to this, a categorization intended to be used as an entry point for a syntactic parser software should contain sufficiently precise tags in order to avoid incorrect syntactic analyses. Regardless of whether the annotations are made by software or by a human, it may also happen that some useful distinctions cannot be annotated in a reliable way, because they imply that many cases are ambiguous (see section 7.5). In this case, simplification becomes necessary.

While preparing the instructions for the annotation process, it is important for each category to be clearly defined so that the annotators know how to use them, in cases where the annotation is performed by humans and not automatically. For example, for annotating speech acts in a corpus, a tag like *question* could be interpreted very differently depending on the annotators, if no explanation is added. Some will only annotate interrogative syntactic forms as in (8) with this tag, whereas others will also include indirect interrogatives as in (9), or even declarative assertions which in context have an interrogative value as in (10).

(8) *Who will come to the party?*

(9) *I wonder who will come to the party.*

(10) *I am dying to know the guest list.*

In order to avoid these problems, a set of tags should be accompanied by a list of criteria specifying in which contexts each tag should be used (see section 7.6). For example, the category *question* could be defined as follows for an annotation of speech acts: “Any utterance used in context as a request for information, whether a direct or indirect one”. This definition would invite annotators to include examples (8)–(10) in the category. However, the definition of interrogative category in a syntactic annotation would probably be more restrictive, only including interrogative syntactic structures. Structures like (9) and (10) would be classified into the category of declarative sentences.

In sum, the definition of annotation instructions largely depends on the linguistic phenomenon studied. In general, it is preferable to choose an annotation scheme as neutral as possible from a theoretical point of view and, in any case, to stick to categories clearly identified and widely accepted in the literature. The results of the annotation will be easily understood and it will be possible to reuse it in future work. As far as possible, the chosen tag sets should be based on annotation standards which have been broadly accepted in the literature. Should these standards be lacking, it would be a good idea to consider the annotation schemes used in previous studies. We should not rule out the point where innovations are desirable or even necessary, but these need to be clearly justified in relation to existing schemes.

Once the tag set has been defined, the corpus processing phase can begin. The first step is to identify which occurrences will be annotated in the corpus. For some phenomena such as the annotation of morphosyntactic categories or speech acts, every word or sentence in the corpus will be involved. However, for other phenomena, the annotation will only refer to very precise elements in the corpus. For example, an annotation of discourse markers such as *well*, *you know* and *I mean* will only relate to these elements.

Annotations covering the entire corpus are often made using computing tools, which we will describe in the next section. In the rest of this section, we will focus on a detailed annotation of specific elements in the corpus. For these annotations, a strategy should be deployed to identify relevant elements. In most cases, identification must be done on raw data, so – whenever possible – it is important to find one or more lexical elements that can be automatically identified by a concordancer. For example, discourse markers can be identified by looking for lexical elements encoding them as *you know*, *I mean*, etc. An annotation of cleft structures in French can start by looking up structures containing the verb form *c'est*. We have offered examples of this type of research in Chapter 2. In other cases, such research will benefit from a preliminary part-of-speech tagging or even from a parsing analysis performed automatically. For example, in order to study causal relations in French, part-of-speech tagging makes it possible to only look for occurrences of *car* working as conjunctions and eliminating those which are nouns (a type of vehicle).

Whatever the strategy used for retrieving data, the results obtained will in most cases require validation and manual sorting in order to eliminate irrelevant occurrences. For example, occurrences of expressions like *you know* and *I mean* do not always correspond to a discourse markers. Sometimes, these expressions are also used in a compositional manner, as in (11) and (12). The two examples have been drawn from the work of Frances Hogdson Burnett *The Secret Garden*, included in the children's literature *ByCoGliJe* corpus (see Chapter 5).

> (11) "*You ought to show me the door today: but I don't believe you know!*"

> (12) *"Perhaps it means just what I mean when I want to shout out that I am grateful to the Magic".*

To eliminate these occurrences, data must be assessed and sorted manually. This type of processing can be carried out effectively by exporting the research results performed with a concordancer into a spreadsheet and by processing every occurrence on a specific row. The rows containing parasitic occurrences can be easily set aside by annotating them with a tag describing an "irrelevant" occurrence. Thanks to this tag, such occurrences can then be separated from other data. This strategy is preferable to purely and simply erasing them, at least at first, since this makes it possible to easily retrieve them when necessary.

Once the corpus to be annotated contains only the relevant occurrences, the annotation process itself can then begin. Whatever the annotation considered and the tag set chosen, the annotation of the first occurrences is generally difficult and many problems and borderline cases arise. It is often impossible to anticipate which dubious cases will appear since corpus data are always much more complex and ambiguous than the reference sentences found in dictionaries. Let us take an example to illustrate this difficulty. According to the Lexconn database (Roze *et al.* 2012), the connective *dans la mesure où* in French can be used for communicating two different discourse relations in context. First, a condition relation, as in (13) and second, a causal relation, as in (14). These two sentences represent invented examples.

> (13) *Je suis d'accord de le voir,* ***dans la mesure où*** *il vient seul.* (I agree to see him, **provided that** he comes alone.)

> (14) *Il n'est pas étudiant,* ***dans la mesure où*** *il n'est pas inscrit au cours.* (He is not a student, **as** he has not enrolled for the course.)

These discourse relations are much more difficult to identify with certainty in examples (15)–(17), drawn from the *Europarl* corpus (see Chapter 5).

> (15) *Tel que je le connais – et je le connais bien – je lui fais confiance: ce mouvement ne va certainement pas s'arrêter et,* ***dans la mesure où*** *il ne s'arrêtera pas, il sera conduit avec habilité, M. Pujol et les autres ministres-présidents n'en manquent pas.* (As much I know him – and I know him well – I trust him: this movement will certainly not stop and, **as long as** it does not stop, it will be led with skill, Mr. Pujol and the other Prime Ministers will see to it.

> (16) *Mes chers collègues, cette énumération résulte de vos propres interventions.* ***Dans la mesure où*** *on a même demandé l'insertion des familles "homosexuelles", je ne vois pas pourquoi vous protestez.* (My dear colleagues, this list results from your own speeches. **Insofar as** we have even asked for the insertion of "homosexual" families, I don't see why you should protest.)

(17) ***Dans la mesure où*** *nous nous dirigeons vers l'adoption d'un nouveau traité, qui plus est constitutionnel, le dessein d'une Europe élargie et sa finalité doivent être mis en débat dans nos sociétés, sans quoi nous ne serions pas à l'abri d'un incident majeur.* (**Inasmuch as** we are moving towards the adoption of a new treaty, which is a constitutional one furthermore, the aim of an enlarged Europe and its purpose must be debated in our societies, without which we would not be immune from major incident.)

These examples illustrate the complexity and the ambiguity of the real corpus data when compared to invented examples. For example, in (17), we may wonder whether the speaker takes it for granted that Europe is moving towards a new treaty (and therefore presents it as a cause) or whether, on the contrary, it is a condition for a treaty to be adopted.

These difficult cases must be resolved by following systematic criteria guiding all decisions. For example, in (17), the use of the present continuous rather than the conditional in the segment following the connective seems to indicate a certain degree of certainty, which is more compatible with a causal interpretation than a conditional one. Thus, verbal tenses can be used as one of the clues that will guide classifications. The adoption of these criteria does not grant that the annotation will be unbiased. On the contrary, since every annotation is based on some kind of interpretation on the part of the annotator, such biases are impossible to avoid. These biases can nonetheless be minimized by having several annotators carry out the same annotation. Furthermore, the adoption of systematically applied explicit criteria ensures the consistency of the annotation and later helps the rest of the scientific community to assess it, or even to criticize it (see Spooren and Degand (2010) for an in-depth discussion of this problem).

Annotation criteria can only be fully determined on the basis of the cases encountered in the corpus, since an exhaustive list of problems is often difficult to anticipate. This is why it is wise to test the categories to be annotated on a small portion of occurrences, ranging, for example, from 50 to 100 depending on the difficulty of the annotation scheme and the total number of occurrences to annotate, then refine the criteria or even redesign the categories on the basis of this first annotation. For example, if an initial distinction proves impossible to discern sensibly in the corpus by the annotators, it should be abandoned. Conversely, if a category seems to cover too many disparate cases, it will have to be refined. Annotation should therefore start from theoretical criteria, but will be progressively modified depending on the nature of the data to be annotated, as illustrated in Figure 7.2.

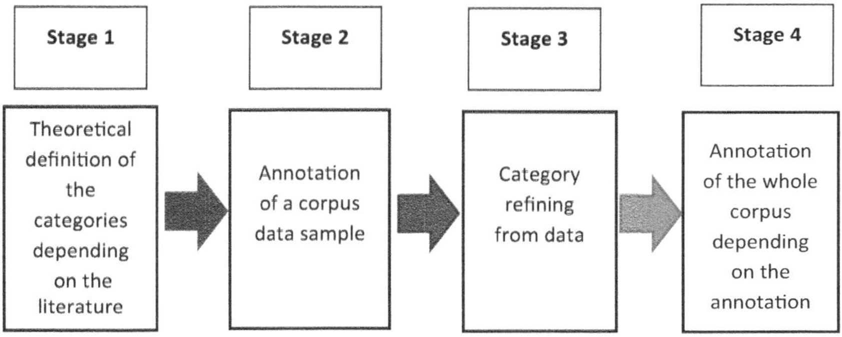

Figure 7.2. *Stages of the annotation process. For a color version of this figure, see www.iste.co.uk/zufferery/corpus.zip*

7.5. Annotation tools

We can classify annotation tools depending on the automatic or manual processing that they bring into play. On the one hand, there are the tools making it possible to carry out annotations in an automatic way, for example, by means of part-of-speech tagging or parsing. On the other hand, there are tools that facilitate the process of manual annotation of data by providing an interface to perform the annotation, as well as a format for representing such annotations. In this section, we will briefly introduce some of these tools. We should beware that there are a very large number of them and that their more or less suitable character depends on the type of annotation the researcher has in mind. It will therefore be up to each researcher to identify the tool best suited to his corpus and his/her research question. Very often, a good starting point is to identify which tools have been used in similar studies in the literature.

The usefulness of part-of-speech tagging is such for corpus studies that this annotation is now directly embedded into some corpus creation tools. The Sketch Engine platform, which we presented in Chapter 6, automatically performs this tagging process when a new corpus is created and this can be done for several languages (see below). Since this platform enables you to create new corpora both from the Internet and from manually inserted files, it is very convenient to use it for carrying out part-of-speech tagging. The set of part-of-speech tags used by this program is presented in the documentation provided on the site. Let us also note that systems for annotating part-of-speech tags (*POS taggers*) should be carefully developed for each language and that the sets of tags corresponding to grammatical categories often vary from one language to another (in concrete terms, their notation may vary even more). However, there are also initiatives there to build up sets of grammatical categories, which if not identical, are at least compatible between languages in terms of fundamental categories (see, for example, the Universal Dependencies project). The tags used may also vary depending on the corpora consulted in different languages, since the annotated corpora available on the Sketch Engine have not all been annotated in the same way. The list of tags used for a corpus is presented on the site. Let us also remember that due to these differences between languages, some of the tools offered online may only work for English. However, Sketch Engine makes it possible to tag a corpus in French as well as in many other languages. If necessary, this annotation can be corrected manually.

Once annotated, the elements of the corpus can be searched by their POS tags by means of a concordancer. For example, Sketch Engine provides the option to search by lemma or by grammatical category. In order to combine search criteria, the CQL query option (*Cassandra Query Language*) must be chosen. For example, to look for all the adjectives that are used with a form of the word *acteur*, the CQL query should take the following form: [lemma = "*acteur*"] [tag = "ADJ"]. This research specifies that all forms of the lemma *acteur* will be retrieved from the corpus when a word tagged as an adjective appears to its right. To find the adjectives used to the left of this word, the element order of the query must be reversed [tag = "ADJ"] [lemma = "*acteur*"]. The CQL syntax needed for formulating queries is documented in the search interface.

The annotation of syntactic dependencies can also be done automatically using computer tools, but so far these have not generally been included in the interfaces for creating corpora, and their use requires natural language processing (NLP) skills that go beyond this book. Interested readers will be able to use toolbox components designed for NLP such as GATE, NLTK or spaCy to make these annotations (see URLs at the end of the chapter).

Manual annotations can be done using tools that help processing and reusing annotations, without replacing the effort of human reflection when categorizing each occurrence. For example, Brat is an online tool that makes it possible to annotate not only entities in a corpus, but also the relations between them. We can thus annotate events described in a corpus, as well as the links between the various participants in these events. For example, a verb like *invite* in (18) implies an argument structure requiring the annotation of the agent making the invitation (in this case, Max) and the invitation's recipient (Lili). The sentence also specifies other characteristics about the event, namely, the place (at his house) and the reason (for a meal), in the form of adjuncts at the end of the sentence.

(18) Max invited Lili to his house for a meal.

All of these are part of the event (*invite*) and can be linked to it by an annotation describing the named entities involved and the semantic links between them. Being able to annotate relations is also essential for associating anaphoric relations in a text. The annotations made with Brat can later be used for research in the Sketch Engine.

Another example is the EXMARaLDA tool, which has been specifically developed to assist in the transcription and annotation of spoken corpora, and later be able to manage them. This tool makes it possible to perform a time-aligned annotation with audio or video files, to insert metadata and to have access to the results of the annotation in different formats.

Many tools, among which we can mention EXMARaLDA, use the XML format, which is a coding language (*eXtensible Markup Language*) for defining, and then storing and transmitting structured data. In this language, the elements of a text are marked up using named tags including one or more attributes. These elements can be embedded into each other depending on the needs of the structure.

For example, an element like *word* can be embedded into a *sentence* type of element. Taggers in XML thus provide a good way for associating annotations with corpus data. However, in order to be interpreted, a document marked up in XML coding must follow certain syntax, in the form of a document type definition (DTD), which specifies the list of tags with their potential attributes, as well as the possible order of appearance for tags. The most widely used XML schema for coding corpus data is the one provided by the TEI (*Text Encoding Initiative*). XML can also be used for encoding the metadata of a corpus (see Chapter 6, section 6.4). In those cases standards like the Dublin Core can be used. Tools have been developed in order to make data annotation in XML format easier, such as the Nite XML tool, for example.

For annotations of a specific linguistic phenomenon (in the situation presented at the end of section 7.2), one solution would be to retrieve the relevant data from the corpus, and then to annotate them separately. For example, we can retrieve all the discourse markers using a concordancer, and then annotate their function in a file. Concordancers generally make it possible to export the data retrieved in text format. However, it is afterwards essential to import the data into spreadsheet software rather than to annotate them in a simple word processor. Actually, a spreadsheet makes it possible to view the data as one occurrence per row in a single column. The following columns will be used for inserting the annotations. Thanks to the functionalities offered by the spreadsheet, the different annotations can then be counted, compared by means of a crosstab and whenever necessary, they can be inserted into statistical analysis software (see Chapter 8, section 8.6).

7.6. Measuring the quality and reliability of an annotation

The notion of quality of an annotation is particularly useful for annotations made automatically, such as part-of-speech tagging. In this case, the quality of the automatic annotation is measured in comparison with a reference annotation produced by human annotators. This annotation is presumed to be completely correct, that is, it matches what has been deemed appropriate in the annotation scheme. It was by means of this type of comparison that the annotations offered for the *Google Books* corpus in section 7.2 were considered as suitable tagging and parsing systems.

This definition of the *accuracy* of an annotation is often subdivided into two separate criteria. On the one hand, the *recall* measures the number of occurrences of each category correctly found by the system. For example, we can report the number of noun occurrences in the reference annotation that the system correctly identified as such, weighing the total number of nouns in the reference. On the other hand, *precision* measures the number of occurrences properly tagged as nouns, from among all the ones tagged by the system as such. For every category, we can combine the recall and precision scores by calculating their harmonic mean (called F1 score), with identical or different weights. Finally, we can express the overall quality of the automatic annotation by calculating a mean of F1 scores per category, by weighing every category according to its number of occurrences in the reference (micro versus macro mean).

The reason for considering recall and precision at the same time (e.g. with the F1 score) is as follows, illustrated with the example of noun tagging. If a system had a strategy for annotating all ambiguous words as nouns (e.g. *rain*, *break* or *close*), its recall would certainly be excellent, but its precision would be very low, since the tag *noun* would probably be mistakenly assigned in many cases. Conversely, if a system annotates as *noun* only the elements that are preceded by a determiner, its precision would be much better this time, since the number of wrong taggers would be lower. However, its recall would be weak because it would fail to correctly tag all the proper nouns. Overall, part-of-speech taggers currently produce results with more than 95% accuracy in the most studied languages such as English and French. Residual errors are either corrected manually or in the case of very large corpora such as the *Google Books* corpus, they are left as such, because their prevalence is low enough not to significantly bias the results of a research.

While calibrating the performance of automatic tools is ultimately based on a reference human annotation, the question arises whether the annotations made by a human are necessarily 100% accurate. Most of the time, this is not the case. As we have seen, annotation always requires a certain type of interpretation and this may vary slightly from one annotator to another. This is why it is crucial to have annotations carried out by several annotators and to calculate the degree of agreement among them, as we will see below. In addition, humans can also make mistakes due to fatigue, lack of attention or a poor understanding of the annotation instructions. Finally, some examples are really ambiguous because the boundaries between certain categories may be blurred. To understand the ambiguity of such examples, one possibility is to enable the use of two tags simultaneously. The examples annotated in that way can later be treated separately in data analysis and can offer new insights on the nature of the categories thus defined.

In order to improve the reliability of annotations made by humans and to help define clear categories, a commonly used method is to have the same annotation made by two different annotators. The more convergent their annotation, the more it can be considered reliable and the divergent case can be resolved through a discussion *a posteriori*. Furthermore, this double annotation process will make it possible to indicate whether the categories have been poorly defined. As a matter of fact, if a certain annotation turns out to be impossible to achieve in a convergent manner for human annotators, this indicates that the phenomenon may be poorly defined or that the categories have been poorly delimited.

In order to measure the agreement between two annotators, a first solution is to count the number of times they choose the same tag and to deduce a percentage of convergent annotations. To do this, the two annotations must be compared as in Table 7.1.

As an example, this table summarizes the annotation of the two possible connective functions for *dans la mesure où* we mentioned above, carried out by two annotators. Numbers indicate that the two annotators agree that the connective communicates a condition relation in 20 out of the 100 sentences they examined, and a causal relation for 60 other sentences. Conversely, there are 14 sentences where annotator 1 considered that the connective indicated a cause, whereas annotator 2 thought it indicated a condition, and 6 sentences in the opposite case (thus, the total of the sentences in the four situations is 100). In total, the annotators agreed in 80% of the cases.

		Annotator 1	
		Condition	**Cause**
Annotator 2	**Condition**	20	14
	Cause	6	60

Table 7.1. *Cross-tabulation of the results of a double annotation*

The problem with this estimate is that it does not indicate whether this agreement should be judged as satisfactory or not. What is more, it does not take into account the probability of an agreement which may be the result of chance in some cases. To illustrate this problem, let us imagine that a first annotator always chooses the tag *condition* to annotate data that actually has 50% condition relations and 50% causal relations. His agreement with a second annotator who would actually analyze the sentences and annotate them correctly would still be 50%. If, however, the real proportion of condition relations in the corpus is 80%, the agreement obtained by chance would be 80%!

A better approach is to use the *kappa* coefficient (Cohen 1960), used for measuring the agreement between two annotators on a binary classification task by taking into account not only the agreement observed, but also an estimate of the probability of agreement which can be obtained by chance. This last point is of great importance. Indeed, an observed agreement of 90% does not have the same meaning if the probability of chance agreement is 50% or 80%. However, this probability varies depending on the proportion of the two classes and is minimal when the two classes have the same frequency. So, unlike a simple count, agreement measured through the *kappa* coefficient acknowledges this distinction. The formula for calculating the *kappa* coefficient (k) is as follows:

$$K = \frac{P(O) - P(E)}{1 - P(E)}$$

In this formula, P(O) corresponds to the proportion of agreement observed during the classification task and P(E) corresponds to the statistical expectation of agreement. The value of P(O) is obtained by dividing the number of matching responses (classifications) by the total number of responses. The value of P(E) is obtained by calculating probabilities, by estimating the average of concordant classifications when the proportion of each class is fixed to the value observed for each annotator. Thus, if each annotator divides their responses into two classes of comparable sizes, P(E) will be close to 50%, whereas if the sizes are unbalanced in the two annotators, P(E) will also increase, since they will have more chances of annotating the most frequent class at the same time[1].

The *kappa* coefficient may have a value oscillating between -1 and 1. A coefficient equal to zero means that agreement between the two annotators does not exceed the one obtained by chance. A negative value indicates an opposite correlation, obtained when annotators disagree the most. The level of agreement is then lower than the one which could be obtained by chance and the value of P(O) is close to 0. Conversely, the maximum value K = 1 indicates that annotators always agree, the value of P(O) then being 1. So, in practice (for a task that is somewhat coherent), the *kappa* coefficient

1 If p1 and p2 are the observed class proportions for each of the annotators 1 and 2, we can say that $P(E) = p1*p2 + (1-p1)*(1-p2)$.

generally varies between 0 and 1, that is, between results that can be obtained by chance and total agreement. This does not mean, however, that any coefficient above zero is a good result. According to Krippendorff (1980), a result greater than or equal to 0.8 reflects a reliable agreement, whereas a result between 0.67 and 0.8 makes it possible to establish the probable presence of agreement.

The *kappa* coefficient can be calculated automatically on online statistical calculation sites like VassarStats, which we will discuss in further detail in Chapter 8. In the case we presented in Table 7.1, the calculated *kappa* coefficient is 0.53. Thus, despite the apparently high proportion of agreement equal to 80%, this score indicates that agreement between the annotators is not reliable and should be improved, in particular by re-discussing cases of disagreement in order to adopt a strategy for processing them in a convergent and systematic way. Naturally, it is not enough to resolve disagreements *a posteriori*, but is better to clarify the annotation instructions so that, on a new set independently annotated by each of the annotators, the *kappa* coefficient has values indicating acceptable or strong agreement.

Finally, we should point out that for some research projects, it is not possible to engage two annotators for the whole annotation task. Several solutions are then possible. First, to double annotate a small part of the data to verify that the scheme can be applied convergently, and then to have the rest of the data annotated by one person. Another method is to test the agreement of an annotator with themselves over time. For this, it is possible to re-annotate the same portion of data after a certain time interval, in order to measure the degree of convergence between the two annotations.

7.7. Sharing your annotations

Given the importance of the effort invested in an annotation process, many researchers choose to share their annotations, which may benefit other researchers. As we have seen in this chapter, in order to make sharing and reusing of annotations easier, it is important to use categories that are as theoretically neutral as possible. It is also important to be able to export annotations in a standardized format, based on the XML language, for example.

Sharing annotated data also implies that the annotation process and the categories used are clearly documented in an annotation manual, which will be provided to future users together with the data. This manual should allow them to understand and to reuse annotations. For this to be possible, such a manual should include two types of information. First, a list of the tags used with their definition, in the way of a mini-glossary. For example, such a list for an annotation of speech acts could take the form illustrated in (19), whereas a list of POS tags could take the form shown in (20).

(19) *Req_dir:* direct request

Req_ind: indirect request

Que_dir: direct question

(20) *ADJM1:* masculine singular adjective

ADJM2: masculine plural adjective

ADJF1: feminine singular adjective

In addition to the tag glossary used, the manual should explicitly document the way in which these tags have been defined and used in the corpus by means of examples. For instance, the definition could be as follows in the case of indirect requests:

> "The *direct request* tag will be used for all statements containing a form of request addressed to the interlocutor and containing either a verb in the imperative, or a performative verb such as *ask* or *order* in the first-person present tense, active voice."

The annotation manual should also list the rules that have been applied to certain borderline or ambiguous cases in the corpus in order to deal with them systematically. It should also provide information on the corpus processing that preceded the annotation process. For example, it should specify the manner in which the corpus has been segmented into words, sentences, utterances or discourse segments. A form of segmentation is actually a necessary step for any analysis relating to a certain linguistic level and such a segmentation process also poses difficulties which must be resolved. For example, does segmentation into words also include compound words? Does the definition of discourse segments include sentences without verbs? As we saw at the beginning of the chapter, these decisions influence the annotation process and must be clearly documented. Finally, when automatic processing tools have been used for preparing and (helping to) annotating the data in the corpus, they must be clearly indicated. Likewise, revisions that have been made at different stages of the annotation process must be documented, as well as successive versions of the corpus that have been produced, if applicable. Many examples of annotation manuals can be found online and are provided with all the corpora made available to the public.

7.8. Conclusion

In this chapter, we introduced the notion of linguistic annotations and discussed the different ways in which these annotations can be incorporated into a corpus. We first reviewed the different types of annotations, covering very diverse elements ranging from phonemes to discourse relations. We then detailed the different stages that make up an annotation process and stressed the importance of good methodological practices, so that the annotation is as valid and reusable whenever possible. We then briefly presented some tools that let you to automatically make annotations or guide manual annotations. We have also seen that these tools often work in standardized formats like XML, which makes it easy to share annotations. We have also addressed the problem of the quality and reliability of annotations and argued that in the case of human annotations, it is difficult to define a quality standard, since each annotator partially interprets the data while annotating them. However, an annotation can be tested from the point of view of its reliability. An annotation is reliable if two annotators produce convergent annotations or if the same annotator produces convergent annotations at two different times. Finally, we presented some recommendations for the creation of an annotation manual, which documents both the content and the annotation process itself in order to enable other researchers to reuse it.

7.9. Revision questions and answer key

7.9.1. *Questions*

1) Explain three advantages of annotating corpus data.

2) Imagine three examples of corpus studies for which part-of-speech tagging is necessary.

3) In the *Frantext* corpus powered by Sketch Engine, look for the nouns *orange* and *nage*, using annotation into grammatical categories. How many unwanted occurrences (adjectives and verbs) does this method allow you to eliminate?

4) Go back to the French cinema corpus that you created in Chapter 6 using Sketch Engine. Use the CQL option to search for adjectives that are used with the words *actrice* and *film*.

5) Using the AntConc, retrieve the occurrences of the connective "*dans la mesure où*" in the *Advanced Literacy* corpus available on Ortolang (see Chapter 5). Export the first 20 occurrences to a spreadsheet and annotate them with the "cause" or "condition" tags. Ask a second person to make this annotation and calculate the percentage of agreement and the *kappa* coefficient. If there is no second annotator, annotate occurrences a second time a few days later.

6) Create a mini manual to document the annotations made in 5. What should it contain?

7.9.2. *Answer key*

1) The first advantage of annotated corpora is that they make it possible to answer many more research questions than raw corpora. Indeed, whenever the words to be looked up are ambiguous or the phenomenon to be investigated cannot be approached, by surface features such as words, raw corpora reach their limits. Annotated corpora, however, make it possible to look for elements in the corpus related to syntactic or semantic annotations, for example. The second advantage of annotated corpora is that annotations can be reused later and thus add value to a corpus. This advantage is all the more evident inasmuch the same annotation can often be used for answering different research questions. Finally, the annotation of a corpus represents a transparent process when it is well documented, which enables an objective evaluation of the data by the scientific community, contrary to other methodological choices which are often less transparent.

2) Part-of-speech tagging is useful for almost all corpus studies, be they lexicon, syntax, semantics, pragmatics questions, etc. A first example in relation to lexicon concerns the study of all polysemic words, for which

part-of-speech tagging makes it possible to sort a good part of the irrelevant occurrences (e.g. to differentiate the adjectival uses of past participles). Since most of the words in the non-specialized lexicon are polysemic, this annotation is essential. In the field of syntax, every analysis should start at classifying words into grammatical categories. It is precisely from these categories that syntactic trees are created. Finally, a search for speech acts communicated by performative verbs such as *ask*, *request* and *promise* can be made easier through the use of POS taggers (as well as lemmatization), since it makes it possible to perform searches for first-person occurrences of the present, which are used for communicating performatives.

3) In Frantext, looking for the noun *orange* yields a result of 139 occurrences, whereas the same search but without specifying the grammatical category results in 174 occurrences. Searching by category therefore makes it possible to eliminate 35 adjectival occurrences, or 20% of errors. The noun *nage* results in 154 occurrences, compared to 252 occurrences if classification into categories is not performed. This research therefore eliminates the 98 verbal occurrences, corresponding to 39% of errors.

4) In order to look for the adjectives that are used with the word *actrice*, it is necessary to make a search for occurrences to the right and another one to the left. The syntax for these queries is: [word = "*actrice*"] [tag = "ADJ"]. This research focuses on the word rather than on the lemma in order to exclude the masculine occurrences, *acteur*. The adjectives to the right mainly include nationalities such as *française*, *américaine*, *italienne*, *australienne* and *allemande*. The search for adjectives to the left of the word *actrice* must use the following syntax [tag = "ADJ"] [word = "*actrice*"]. These adjectives are narrower and list modifiers like *meilleure*, *grande* and *pire*. For the word *film*, the search can focus on the lemma rather than on the word, in order to identify singulars and plurals. The syntax is therefore: [lemma = "*film*"] [tag = "ADJ"]. To the right, the results are varied and include adjectives such as *super*, *romantique*, *dramatique*, *important*, *érotique* and *indépendent*. The search for adjectives to the left, with the syntax [tag = "ADJ"] [lemma = "film"] results in the adjectives *meilleur*, *prochain*, *nouveau* and *seul*.

5) Here in the table below is a list of 20 occurrences drawn from the corpus with two possible annotations.

Occurrence in French	English version	Annotator 1	Annotator 2
Dans ce contexte, ils ne pensent pas être de véritables auteurs, *dans la mesure où* les professeurs leur dictent d'une certaine manière ce qu'ils doivent faire et écrire.	In this context, they do not consider themselves the real authors, insofar as the teachers dictate to them what they have to do and write in a certain manner.	Cause	Cause
L'écriture serait pour l'enfant une démarche d'auto-socio-création (terme emprunté au GFEN) *dans la mesure où* cela permettrait à l'individu de développer son potentiel imaginaire par la comparaison de ses représentations mentales avec celles des autres.	For the child, writing would be a process of self-socio-creation (term borrowed from GFEN), insofar as it would help the individual to develop his imaginary potential by comparing his mental representations with those of others.	Condition	Cause
Pour écrire un texte, l'élève doit posséder une certaine connaissance des genres littéraires et se les approprier, *dans la mesure où* les propos s'inscrivent dans des formes précises relevant des genres standardisés plus ou moins adaptables.	To write a text, the student must have certain knowledge of literary genres and has to own them, insofar as the words are inscribed in precise forms which belong to standardized genres, more or less adaptable.	Cause	Cause
Dufaÿs rappelle qu'écrire n'est pas pure création *dans la mesure où* l'auteur reprend une structure et des éléments déjà utilisés et repris par d'autres auteurs.	Dufaÿs recalls that writing is not pure creation insofar as the author picks on an existing structure and elements which have already been used and taken up by other authors.	Condition	Cause
Contrairement au texte de Maurice Carême où le modèle reste fixe, celui de Roger Judenne permettait une plus grande liberté d'expression et laissait une plus grande part à l'imagination *dans la mesure où* les enfants devaient réinventer une histoire.	Unlike Maurice Carême's text, where the model is fixed, that of Roger Judenne enabled greater freedom of expression and left a greater part of the imagination to the point that children had to reinvent a story.	Cause	Cause
De plus, cet exercice plaisait à l'enseignante *dans la mesure où* il apportait de nombreuses valeurs ajoutées pédagogiques.	In addition, this exercise pleased the teacher insofar as it brought many educational added values.	Cause	Cause

L'échange d'idées favorise la création des enfants *dans* la *mesure où* chacun apporte une idée personnelle à laquelle il est attaché et pour laquelle il éprouve une certaine fierté.	The exchange of ideas promotes children's creations insofar as everyone contributes a personal idea dear to him and for which he feels a certain pride.	Cause	Condition
Cet écrit est original *dans la mesure où* l'enfant s'approprie le personnage de Perla, l'ara bleu du film d'animation Rio, ainsi que l'un des lieux du film, le bois de Rio.	This writing is original insofar as the child owns the character of Perla, the blue macaw of the animated film Rio, as well as one of the places of the film, Rio's forests.	Cause	Cause
La consigne est respectée, mais Adam se démarque de ses camarades et de l'auteur *dans la mesure où* l'enfant et l'animal ne restent ni ensemble ni amis à la fin.	The instructions were followed, but Adam stood out from his comrades and the author insofar as the child and the animal did not stay together or friends at the end.	Cause	Cause
Cet exercice semble donc favoriser le statut d'auteur de l'élève, *dans la mesure où* celui-ci est libre de choisir ses personnages et les événements de son récit.	This exercise seems to favor the student's author status, insofar as the latter is free to choose his characters and the events of his story.	Cause	Condition
Il est nécessaire d'en passer par ce genre d'exercice, *dans la mesure où* celui-ci est important dans l'apprentissage littéraire des élèves.	It is necessary to go through this kind of exercise, insofar as it is important for students' literary learning.	Condition	Cause
Le travail de groupe est un choix pédagogique difficile dans *la mesure où* l'enseignante a dû faire attention à ce qu'aucun élève ne soit laissé pour compte au sein du groupe.	Group work is a difficult pedagogical choice insofar as the teacher had to take care that no student was left behind in the group.	Cause	Cause
La contrainte peut certes freiner dans certains cas l'imagination, mais elle est utile *dans la mesure où* elle permet une progression, puisqu'elle place le travail de l'écriture au premier plan, avant l'inspiration réelle.	In some cases, the constraint can certainly slow down the imagination, but it is useful insofar as it favors progression, since it sets the work of writing in the foreground, before actual inspiration.	Condition	Cause

En effet, l'imagination créatrice apparaît comme synonyme d'audace ou de tentation, *dans la mesure où* elle embellit le quotidien et entretient l'espérance.	Indeed, the creative imagination appears as synonymous with daring or temptation, insofar as it embellishes everyday life and maintains hope.	Cause	Cause
Il y a une impossibilité de l'approfondissement de l'imaginaire, *dans la mesure où* l'école attend de l'élève qu'il évolue dans une certaine direction, au niveau du langage tout autant qu'au niveau psychique.	There is an impossibility of deepening of the imaginary, insofar as the school expects the student to evolve in a certain direction, at the language level as well as at the psychic level.	Cause	Cause
Le travail de groupe constitue un choix pédagogique difficile *dans la mesure où* il est compliqué de faire en sorte que chaque élève soit auteur.	Group work is a difficult pedagogical choice since it is complicated to ensure that each student is an author.	Cause	Condition
En conclusion, nous pouvons dire qu'il est possible de considérer que l'élève est auteur *dans la mesure où* les écrits l'impliquent en tant qu'enfant, faisant ainsi appel à ses sentiments, son vécu, mais aussi à des choix d'écriture et à son expérience personnelle de l'écrit.	In conclusion, we can say that it is possible to consider that the student is an author insofar as the writings engage him as a child, thus appealing to his feelings, his experience, but also to writing choices and his personal writing experience.	Condition	Condition
Cette mise à distance de l'évaluation certificative – bien que rien n'interdise d'utiliser le dispositif portfolio précité dans cette optique, *dans la mesure où* les critères d'évaluation élaborés par la recherche sont "des sortes d'idéaux à atteindre" –, favorise certainement l'engagement des étudiants dans la démarche d'écriture réflexive.	This distancing of the certification evaluation - although nothing prohibits the use of the aforementioned portfolio device in this perspective, insofar as the evaluation criteria developed by the research are "kinds of ideals to be attained" – certainly promotes student engagement in the reflective writing process.	Condition	Condition

Ces exemples illustrent des activités constructives *dans la mesure où* l'écrit est suscité pour répondre à une finalité conscientisée par l'élève et l'amènent ainsi à faire le lien entre le message à produire et ce qu'il va écrire.	These examples illustrate constructive activities insofar as the writing is aroused to meet a conscious aim of the student and thus invite him to make the link between the message to be produced and what he is going to write.	Cause	Cause
L'écriture chez le jeune élève peut aussi être motrice de lectures *dans la mesure où* la réponse à un projet d'écriture suscite de nouvelles lectures ciblées.	Writing in young students can also be a source of reading insofar as the response to a writing project generates new targeted readings.	Condition	Condition

In order to be able to measure agreement, we must create a crosstab with the two annotations. In the Excel spreadsheet, this can be done automatically by choosing the data and choosing the pivot table option. In this table, annotator 1 should be recorded on the rows and annotator 2 in the columns. For the annotation above, this table looks as follows:

		Annotator 1	
		Cause	**Condition**
Annotator 2	**Cause**	10	4
	Condition	3	3

The two annotators converge on 13 of the 20 sentences, that is, there is 65% of agreement. This agreement corresponds to a *kappa* coefficient of 0.2. This value is very low and shows that the annotation is not reliable and should be revised.

6) A manual documenting of this annotation should contain the list of tags as *Condition* and *Cause*, together with their definition. It should also contain the rules to follow in ambiguous cases, for example: "The condition relation will only be chosen if the verbal tense used is the conditional tense". This manual should also indicate how the annotation was carried out (by one or two people, throughout how many sessions) and whether successive versions were produced.

7.10. Further reading

McEnery *et al.* (2006) present the different types of linguistic annotations that can be carried out on a corpus. Kuebler and Zinsmeister (2014) is an introductory work devoted to the annotation of a corpus. The current most complete work on linguistic annotations is the one by Ide and Pustejovsky (2017), the first volume of which offers a theoretical overview of a wide range of problems related to the annotation of linguistic data, whereas the second volume presents a series of major projects of annotated corpora in different languages and for different linguistic levels. Ways to measure agreement between annotators are described in detail by Artstein (2017). Finally, Spooren and Degand (2010) discuss the difficulties associated with annotation and suggest different solutions to reduce bias.

8

How to Analyze Corpus Data

Quantitative corpus studies produce numerical results such as word frequency or the number of occurrences of a given syntactic structure across different text genres. By using statistical tests, it is possible to analyze these numerical data and reveal tendencies that are not always visible to the naked eye, and to evaluate whether or not the observed trends are statistically significant. The purpose of this chapter is to introduce some simple statistical methods that are commonly used for processing corpus data. To begin with, we will introduce the concept of descriptive statistics. Then, we will present two types of descriptive statistics that, respectively, make it possible to measure lexical diversity (the *type/token* ratio), and to calculate lexical dispersion in a corpus. We will later describe the principles underlying inferential statistics. In addition, we will see that data from corpora correspond to different types of variables, and that such categorization is important in view of deciding which statistical test should be used. Finally, we will illustrate the use of inferential statistics by presenting a commonly used test in corpus linguistics, namely the *chi-square* test, which determines whether frequency differences between categories are significant.

8.1. Descriptive statistics for corpus data

The first step for analyzing quantitative data is to describe the content of a corpus. For example, this step could involve calculating the number of sentences in the passive voice used in a corpus of journal articles. Imagine that, in this study, the passive sentences were retrieved from 10 texts from different newspapers, resulting in the figures reported in Table 8.1 for each text.

Text no.	1	2	3	4	5	6	7	8	9	10
Number of passive sentences	47	58	64	76	78	89	91	96	104	120

Table 8.1. *Total number of passive sentences per text*

In order to summarize the number of occurrences of passive sentences observed in this sample, the most obvious solution is to calculate the mean number of passive sentences observed. In the previous example, the mean equals 82.3 passive sentences per text in the corpus (a total of 823 divided by 10 texts). Mean calculation is a good way of summarizing data as long as the sample is made up of homogeneous values. For instance, if text no. 10 had contained 1,200 passive sentences instead of 120, this extreme would have completely changed the observed mean, from 82.3 to 190.3. In this case, the mean would no longer be truly representative of the data, since it would be higher than all the values observed except for one. The mean is a central tendency indicator; that is, it offers information about the central value of data distribution. At this point, and taking it as an isolated indicator, it is not entirely sufficient for summarizing the data obtained.

In order to complete the information offered by the mean, it is also necessary to observe data dispersion with the help of two indicators: the variance and the standard deviation. If we take a look at Figure 8.1, we will see that every value included in the calculation of the mean is distant from it by a certain amount (shown in red); this is what we call dispersion. If these distances, either positive or negative, are added together, their sum is equal to zero because the distances of the values below the mean are compensated by those located above. In order to quantify the total distance, it is necessary to transform the negative values into positive ones, for example, by squaring them. The sample's variance is calculated on the basis of the square of the mean distances divided by the number of observations minus one. By calculating the square root of the variance, we obtain the sample's standard deviation, which can be considered as an indicator of the mean's deviation in relation to the mean. In the case of the sample reported in Table 8.1, the standard deviation is 22.2. The standard deviation would drop to 355.22 if the number of occurrences in text no. 10 was 1,200, rather than 120. A standard deviation that is higher than most of the numbers found in the corpus immediately provides an alert concerning the data's lack of homogeneity. The standard deviation can be calculated automatically thanks to a dedicated function in spreadsheet software (e.g. Excel). This measurement can also be automatically calculated in software devoted to statistical tests (see URLs at the end of this book).

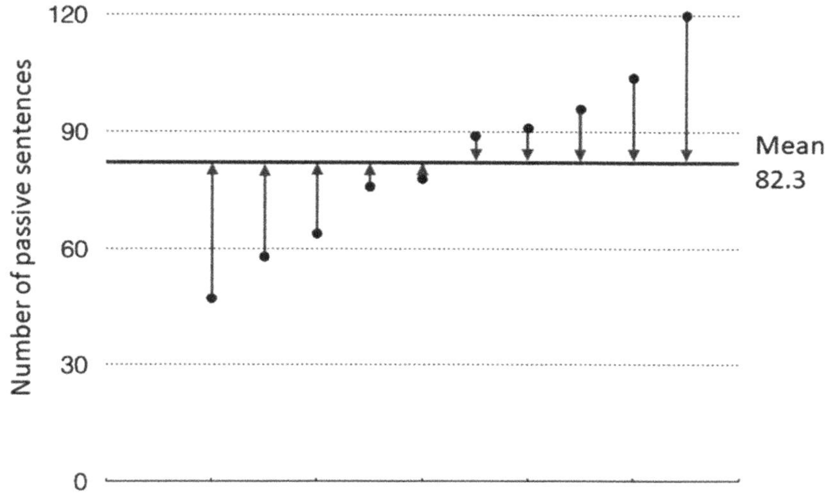

Figure 8.1. *Value dispersion around the mean*

Another measure of central tendency is the median, placed at the middle of the different values found in the corpus, so that the data set is divided into the lower half and the upper half. In the case of our example, which includes 10 values ranked in ascending order, the median is equidistant from the fifth and the sixth value, that is, between 78 and 89. Its value is 83.5, the mean of the two intermediate values. When the number of values is an odd number, the median is simply the value in the middle. What is interesting here is that even if the value of 120 in text no. 10 was replaced by 1,200, the median would be the same as before. We can therefore see that the median is more appropriate than the mean for summarizing data in the presence of extreme values.

Before performing descriptive statistics on the data obtained in corpus studies, it is necessary to make the number of occurrences comparable, and this can be achieved through the use of different sources. For example, in order to compare the number of passive sentences in the above-mentioned

10 texts, it would have been necessary to ensure that all the texts had a comparable number of words. Indeed, we can observe that text no. 10 contains almost three times more passive sentences than text no. 1. If text no. 10 were roughly three times longer than text no. 1, the two texts would be equivalent in terms of the proportion of passive sentences. If this were not the case, then the two texts would be different. As a result, when the sources are of variable length, which is generally the case, it is necessary to transform the number of occurrences into relative frequencies, so that they can be easily compared. Let us take another example to illustrate this point. Imagine a study aiming to compare the use of passive sentences in spoken and written modes, and finding 67 occurrences of passive sentences in the spoken corpus versus 3,372 in the written corpus. If these figures come from corpora of different sizes, for example a written corpus of 3,179,546 words and a spoken corpus of 573,484 words, they cannot be compared directly. As a matter of fact, a larger corpus increases the probability of finding more occurrences of a phenomenon. Thus, in order to determine whether passive sentences are used more frequently in the written than in the spoken form, it is necessary to transform the data according to the same base of normalization, for example the number of occurrences per 10,000 words, per 100,000 words or even per million words. To turn a number of occurrences into a relative frequency, we need to apply a rule of three, by dividing the number of occurrences found in the corpus by the total number of words in the corpus, then multiplying by the base of normalization, as shown in the example below, which has a base of normalization equal to 10,000. This (fictitious) comparison indicates that passive sentences are in fact almost 10 times more frequent in written than in spoken discourse:

$$\frac{67}{573{,}848} \times 10{,}000 = 1.17 \quad \frac{3{,}372}{3{,}179{,}546} \times 10{,}000 = 10.61$$

Normalization by means of relative frequencies makes it possible to compare data from different corpora. The choice of the appropriate base of normalization depends on the size of the two corpora. For small corpora of a few tens of thousands of words, a base of normalization set at 1 million words would not be appropriate. If a word appears five times in a corpus of 15,000 words, it would not be wise to conclude that this word would have a frequency of 500 occurrences in a corpus of 1,500,000 words. In fact, the behavior of rare words varies a great deal according to the genre, and even from author to author. For this reason, extrapolating a frequency, obtained on the basis of a small corpus, to a much larger scale does not offer a correct representation of what was actually observed. In general, it is appropriate to choose a base of normalization close to the size of the smallest corpus examined. However, even if it is necessary to normalize data in order to be able to compare corpora, such normalization cannot completely replace raw data, which should also be reported in order to communicate what has actually been observed. For example, raw data can be presented in brackets in a table next to the relative frequencies, as shown in Table 8.2 for the three causal connectives *car*, *parce que* and *puisque* retrieved from the *Sciences Humaines* corpus (see Chapter 5).

car	*parce que*	*puisque*
3.3 (86)	3.61 (94)	0.99 (26)

Table 8.2. *Relative frequency of causal connectives every 10,000 words in the Sciences Humaines corpus, with absolute frequencies in brackets*

In the text of an article, the two frequency elements can be reported as follows: "There were 94 occurrences of *parce que* in the corpus, corresponding to a relative frequency of 3.61 occurrences every 10,000 words".

Finally, we should observe that, in other cases, normalizing data may involve other methods, such as transformation into percentages. For example, Dupont and Zufferey (2017) compared the different translations of concession connectives *néanmoins* and *toutefois* in English in a corpus of newspaper articles, with the results shown in Table 8.3.

	néanmoins	***toutefois***
but	27	64
however	7	26
no translation	38	59
nevertheless	10	3
nonetheless	2	2
other	21	35
Total	**105**	**189**

Table 8.3. *Translations of "toutefois" and "néanmoins" into English in a journalistic corpus (Dupont and Zufferey 2017)*

In order to compare the distribution of the different translations for every connective, data needs to be transformed, as the total number of occurrences in the corpus is different in the two cases. A good solution would be to translate information into percentages, since it makes it possible to quickly compare data distribution across the different categories. This transformation can be done by dividing the number of occurrences in a category by the total number of occurrences of this element, and by multiplying the result by 100. Thus, looking at raw data, we may have the impression that there are more untranslated occurrences for *toutefois* (59) than for *néanmoins* (38). When transformed into percentages, however, these data reveal the opposite tendency, with only 31.2% untranslated occurrences for *toutefois* against 35.2% for *néanmoins*. This result raises the question of whether the difference observed in the corpus reflects a distinction that might be found in all newspaper articles. Let us rephrase it this way: does such a difference enable us to conclude, with a certain degree of confidence, that "*toutefois*" is more often translated into English by a different connective other than "*néanmoins*"? It is exactly this type of question that inferential statistics help us to answer, as we will see later in this chapter. But before that, let us illustrate the use of descriptive statistics with two concrete examples.

8.2. Measuring the lexical richness of a corpus

A useful application of descriptive statistics for corpus data is to calculate the lexical richness of a corpus. In Chapter 7, we saw that the notion of word encompasses different realities, depending on whether we adopt the point of view of morphology (grouping up *lemmas*) or semantics (which focuses on *lexemes*). There is yet another distinction that needs to be applied to the concept of word, since it is fundamental for corpus linguistics. This notion differentiates *word types* from *word occurrences*. For example, think of a teacher asking you to hand in an assignment of "10,000 words". To complete this task, you will count each character string one after the other to verify that you have reached the total requested. For example, according to this definition of "word", there are 12 words in the French sentence in (1). This definition corresponds to what we call word occurrences.

> (1) *La mère de Jacques est plus jeune que la mère de Pierre.* (Jacques' mother is younger than Pierre's mother.)

Other times, however, we refer to a word not to designate the total number of character strings, but the number of *different* words that a corpus contains. For example, if your teacher informs you that you have to learn a vocabulary of 500 words in a foreign language by the end of the year, these are obviously different words. So, according to this second definition of word, there are only nine different words (or word types) in sentence (1), namely: *la*, *mère*, *de*, *Jacques*, *est*, *plus*, *jeune*, *que*, *Pierre*.

Notions such as word type and word occurrence are both very important in corpus linguistics. The size of a corpus is calculated in word occurrences, whereas the number of word types indicates the diversity of the vocabulary used in the corpus. In order to measure the lexical richness of a text, it is customary to calculate the ratio between the number of word types and the number of word occurrences, according to the following formula:

$$\text{type/token ratio} = \frac{\text{number of word types in the corpus}}{\text{number of word occurrences in the corpus}}$$

The greater the type/token ratio, the more lexically diverse the text is. Generally speaking, written genres have a higher ratio than spoken genres. For example, in the spoken corpus of French from French-speaking Switzerland OFROM (see Chapter 5), the type/token ratio is 0.02, whereas in the French section of the children's literature corpus *ByCoGliJe*, it is equal to 0.04.

The problem with the type/token ratio is that it is very sensitive to the size of a corpus (for this reason, the type/token value 9/12 = 0.25 obtained for sentence (1) is in no way representative). The comparison between the two corpora that we have just presented could be reliably established since they are similar in size (around a million words), but this ratio may offer biased results for corpora of very different sizes. Indeed, the larger the corpus, the more the words end up repeating themselves and, therefore, the ratio decreases all the more. For example, the type/token ratio in the *Sciences Humaines* written corpus that contains around 260,000 word occurrences is 0.06, but it drops to 0.007 for the 2012 installment of the *Le Monde* corpus, which has over 26 million word occurrences. We must therefore avoid using this type of measurement on corpora of different sizes. An alternative solution would be to divide the corpora into segments of equivalent size (e.g. 1,000 words), then to measure the ratio for each segment and, after that, the mean corresponding to the different values.

8.3. Measuring lexical dispersion in a corpus

So far, we have discussed how to calculate and report word frequency in a corpus. However, in order to estimate the importance of a word in a corpus, frequency is not the only element to take into account. For example, in a corpus containing scientific articles, the word *linguistics* may appear relatively frequently, due to the fact that a portion of the corpus is devoted to this field. However, this word will probably never be used in other texts in the corpus covering different fields, unlike words such as *analysis*, *hypothesis* or *conclusion*, which will probably be used in all scientific fields. Let us consider another example. In a small corpus of texts produced by 20 French-speaking students, the word *nonobstant* appears more frequently than in other language registers such as journalistic texts, but this higher frequency is due to the propensity of a particular student to use this word very frequently. In this case, it would be inappropriate to conclude that the word *nonobstant* is used more frequently in student texts than in newspaper articles, as this high frequency does not reflect a common practice among students. In order to avoid this type of bias, in addition to word frequency, it can be useful to calculate lexical dispersion in a corpus. This parameter reflects the way in which word occurrences are distributed across the different portions of the corpus.

There are different ways to calculate lexical dispersion in a corpus. One of the simplest ones is to count the number of portions of the corpus in which the word is present. For example, for a corpus made up of texts written by 20 students, if the word *nonobstant* is used by three of them, we can say that its dispersion is 3/20, or that this word is found in 15% of the corpus. Although this measure may be useful in some cases, it is not entirely satisfactory, since it does not reflect the proportion in which this word has been used throughout the three portions of the corpus. If, in this case, one of the students produces 80% of the occurrences and the other two students produce 10% each, it would be inappropriate to conclude that the distribution is homogeneous in the 15% of the corpus where this word appears. For this reason, other, more sophisticated, measures have been developed. Here, we will discuss only one of them, known as *deviation of proportions* (DP), which was proposed by Gries (2008). This measure compares the expected proportion of a word, in the different portions of a corpus, with the real proportion observed. The result produces a number between 0 and 1. The closer the result is to 0, the more homogeneous the distribution in the corpus will be, meaning that the observed distribution differs only slightly from the expected one, if the word were distributed uniformly in the corpus. In fact, if every portion of the corpus has an observed proportion similar to the expected proportion, then the sum of the difference between the two proportions is small.

In order to determine the *expected* proportion, we must count the total number of word occurrences in each portion of the corpus separately, and then divide this number by the total number of word occurrences of the corpus. The proportions *observed* are calculated by taking into account the total number of occurrences of the word, whose distribution we are trying to determine in each portion of the corpus, and dividing it by the total number of occurrences of the word in the corpus.

In order to find the difference in proportions, we then have to subtract the expected proportion from the observed proportion for each portion of the corpus. All of the figures obtained for each portion of the corpus are then added, and divided by 2. The procedure that we have just described is summarized in the following formula, where Σ refers to the sum:

$$Deviation\ of\ proportions = \frac{\Sigma\ absolute\ values\ (observed\ proportions = \text{expected}\ proportions)}{2}$$

Let us work with an example to illustrate this procedure. Suppose we want to know whether the connective *car* was used consistently by the 10 students who compiled a dossier in the *Littéracie avancée* corpus (L2_DOS_SORB sub-corpus). In order to find out, first, it will be necessary to start by counting the number of word occurrences produced by each of the 10 students separately, as in the column "Number of word occurrences" in Table 8.4.

Student	Number of word occurrences	Expected proportion of *car*	Absolute frequency of *car*	Observed proportion of *car*	Difference in proportions
1	2,638	0.098	6	0.167	0.069
2	2,396	0.089	3	0.083	0.006
3	3,452	0.129	3	0.083	0.046
4	4,809	0.179	2	0.056	0.123
5	2,174	0.081	0	0	0.081
6	2,718	0.101	2	0.056	0.045
7	2,017	0.075	4	0.111	0.036
8	2,012	0.075	0	0	0.075
9	1,778	0.066	4	0.111	0.045
10	2,842	0.106	12	0.333	0.227
Total	26,836		36		0.753

Table 8.4. *Data used for calculating the difference in proportions as a dispersion indicator*

Then, divide this number by the total number of word occurrences in the corpus (26,836), so as to obtain the expected proportion of *car* for each student, compared to the total number of occurrences of this word in the corpus. Then, we have to count the number of *car* occurrences produced by each student separately, as in the "Absolute frequency of *car*" column. To obtain the observed proportion, we will then have to divide the number of *car* occurrences per student by the number of *car* occurrences in the corpus (36). The difference in proportions is obtained by subtracting the expected proportion of *car* from its observed proportion (e.g. 0.167–0.098 for student 1). If the result is a negative value, the latter must be transformed into a positive number by changing its sign (as is the case for student 2, for example). In order to obtain the deviation of proportions, it is necessary to add the differences of proportions (0.753) and then divide this result by 2, which gives a final result of 0.38. This proportion is closer to 0 than to 1, which indicates that the distribution of *car* is relatively homogeneous in the corpus.

However, the dispersion measure only becomes truly informative when it is compared to the one obtained for other words. For example, the same calculation indicates that *puisque* has a deviation of proportions equal to 0.45 and *parce que* a deviation of proportions equal to 0.43. *Car* seems to be the connective that is used most consistently by students. The homogeneity of *puisque* and *parce que* is also very similar.

In summary, in addition to frequency indications, it is important to provide information about lexical dispersion in a corpus, either by simply calculating the percentage of texts in which a word is used, or by reporting a dispersion measurement, such as the deviation of proportions we have just described.

8.4. Basics of inferential statistics

So far, we have seen that corpora compile linguistic data collected from texts or recordings. These corpora do not contain all possible linguistic data but represent their samples. The primary goal of research in corpus linguistics is to be able to draw conclusions in relation to a linguistic phenomenon, not only at the level of the observed data sample, but also to generalize it to all data of the same type. To do this, it is necessary to use inferential statistics, a tool that makes it possible to know whether it is correct to generalize the results found, at the level of a sample, to a population.

Before going further in the description of the logic of inferential statistics, we should focus for a moment on the notion of hypothesis, which is the basis for statistics. There are different ways of making hypotheses. We could imagine the following hypotheses in [1] and [2]:

> *[1]* French is a more beautiful language than English.

> *[2]* Children produce few passive sentences.

These hypotheses are interesting but cannot be tested empirically, since it is impossible to collect data that would allow them to be refuted. In fact, hypothesis no. [1] is a subjective value judgment, whereas hypothesis no. [2] has been formulated too vaguely to be empirically testable.

Empirically testable hypotheses must clearly define the variables observed, the relationship between these variables, as well as the measures used for describing them. For example, hypothesis no. [2] could be transformed into no. [3]. Note that, by convention, empirically tested hypotheses are written as H_1.

> *[3]* H_1: Children aged between 5 and 8 years produce fewer passive sentences than those aged between 9 and 12 years.

In an inferential statistics test, it is not research hypotheses like [3] that are tested but alternative assumptions called null hypotheses. A null hypothesis focuses on the opposite fact to the one we wish to prove. The null hypothesis for [3], written as H_0, is [4].

> *[4]* H_0: Passive sentences are used with a similar frequency by children aged between 5 and 8 years and those aged between 9 and 12 years.

The reason why statistical tests assess the null hypothesis and not the research hypothesis itself arises from the philosophical argument that it is not possible to prove that a hypothesis is true in all cases (be they observed or not), whereas it is possible to prove that it is false by presenting a single case when this exists (counterexample). Howell (1998, p. 105) offers the following example. The assertion *Everybody has two arms* cannot be demonstrated by observing only 3,000 people who have two arms (or even by observing all of the people currently alive, as this statement ignores those to come). However, it can easily be refuted by finding a single person who does not have two arms. For this reason, an inferential statistical test aims to measure the chances the null hypothesis has of being true.

An inferential statistics test provides two important values:

– the numerical result of the test, such as the chi-square value (see section 8.5);

– the probability value associated with the result, denoted as p. The value of p can oscillate between 0 and 1. It expresses the probability of the sample data being observed if the null hypothesis were true in the population.

Let us go back to the example of the translations of *néanmoins* and *toutefois*, presented in Table 8.3. In this case, the null hypothesis states that there is generally no difference between the number of times *néanmoins* and *toutefois* remain untranslated, and that the different values observed only result from the natural variation of their proportions in the samples. Indeed, the data collected show that in the sample of texts studied, 35.2% of *néanmoins* occurrences were not translated, nor were 31.2% of *toutefois* occurrences. These values are certainly different, but they could be explained intuitively by some variability in each of the proportions around ideal values, which could be identical for the two connectives. Applying statistical tests makes it possible to determine whether the 4% deviation observed in the samples is due to chance or reflects a genuine difference in translation choices. The interpretation of the test is performed according to the following reasoning: if this probability is high, we cannot exclude the fact that connectives are translated in the same way (the null hypothesis), whereas if it is low, we can say that the difference cannot be the result of chance, and reflects a real difference in translation.

The null hypothesis can be rejected when the value of p is smaller than a certain value. In some fields such as medicine, only a margin of error of 1% is tolerated, and p is only considered significant when it is smaller than 0.01. In other fields, some researchers choose a higher confidence threshold and accept a 10% margin of error (which corresponds to a p value equal to 0.1). In this chapter, we will stick to the 5% (0.05) value, which is the most frequently used in the social sciences.

In summary, the result of a test is said to be significant when p is smaller than 0.05. This amounts to saying that in the case where the null hypothesis is correct, the observed difference (or one greater) might be obtained less than 5% of the time. In the case where the value of p is greater than 0.05, the results obtained do not make it possible to confidently reject this conclusion and the null hypothesis. However, this does not mean that the data prove the null hypothesis (following the logic that which has not been proven to be false must necessarily be true). An insignificant test does not mean that the research hypothesis is false, which in itself would be an interesting result. The only adequate conclusion for an insignificant result is that there are not sufficient elements present in the data to be able to reject the null hypothesis.

Finally, an important distinction that must be made in order to understand inferential statistical tests is the difference between one and two-tailed tests. Two-tailed tests are used when the direction of the difference between two groups or two categories has not been specified by the research hypothesis. For example, if the research hypothesis tested is that the connectives *néanmoins* and *toutefois* are not used with the same frequency in two text genres, but we do not know which type of text is supposed to contain more connectives than the other, then a two-tailed test should be carried out. However, if the way in which the groups differ is specified in the research hypothesis, as in no. [3], stating that older children use more passive sentences than younger children, then a one-tailed test is preferable.

8.5. Typical variables in corpus studies

There are different inferential statistical tests, whose application depends on the type of variable observed. Before describing these tests in depth, we should succinctly refresh the concept of variable and the different forms a variable may take. A variable simply designates something that can display different values. For example, the number of adjectives found in different corpora is a variable. Likewise, the gender of participants in a corpus, or their geographical origin, are also variables. In the literature, types of variable may receive different names. Here, we will follow the approach proposed by Brezina (2018) and draw a distinction between linguistic and explanatory variables. This distinction is very important when applied to corpus data, since corpus linguistics aims to identify the relations between explanatory and linguistic variables.

Linguistic variables are related to frequency elements measured within the corpus, such as the number of adjectives or the number of speech acts like *requests* or *orders*, or the number of negative prefixes such as *un-* or *dis-* found in a corpus, to mention just a few examples.

Explanatory variables are related to the context in which linguistic data are produced. These variables can include gender and geographical region, as we have already seen, and also the textual genre or form (spoken or written discourse), the language in the case of comparable corpora, and the age or language proficiency level in the case of children and learner corpora. In short, these variables gather all the necessary elements for analysis, which are not the linguistic data themselves.

The two categories of variable we have just introduced can be measured through different scales:

– nominal;

– ordinal;

– scalar.

Nominal variables involve values corresponding to different categories, which have no numerical value. For example, a word's grammatical category is a nominal variable that may acquire values such as *noun* or *adjective*, among others. The function we assigned to the connective *dans la mesure où* in Chapter 7 is another example of a nominal variable having two possible values, namely *cause* and *condition*. In these examples, nominal variables fall into the category of linguistic variables. Another example of a nominal variable, this time an explanatory one, could be the region of origin of the speakers of a spoken corpus, for example, Paris, Geneva, Brussels, Montreal. In some cases, nominal variables are coded using numbers, for the sake of simplicity, for example 1 for Paris, 2 for Geneva. However, these numbers are simply category codes and their numerical values have no particular meaning.

Ordinal variables are also used for grouping values into different categories, but they have a natural order. For example, if the participants in a corpus are classified into age groups such as group 1 (18–25 years), group 2 (26–40 years) and group 3 (41–60 years) for sociolinguistic analyses, the *age* variable is an ordinal one. So, here we can see that it is possible to order the different groups, since the participants in group 3 are older than those in group 2, who in turn are older than those in group 1.

In the previous example, while age groups can be ordered as such, it is not possible to do so with the different participants within each group. The variables that can be ordered in this way are called scalar variables, measured on a digital scale, whose points are distributed at equal distance and with a zero point. The property of these variables is that operations such as addition, subtraction, multiplication and division can be performed on them, since they represent measurable quantities, rather than a simple ordered list, as ordinal variables are. For example, if measured continuously rather than in groups, the age of participants can be considered a scalar variable, since the distance between 12 and 13 is the same as between 13 and 14. We can also say that a 28-year-old is four times older than a 7-year-old child. Linguistic variables involving the relative frequency of some kind of phenomenon in a corpus are also scalar variables.

Since different types of variables have different properties, they cannot be processed in the same way. For example, let us imagine that we have collected information concerning the nominal variable *mother tongue* from the speakers in a corpus, and that we have 36 French-speaking people and 40 German speakers. As such, the only type of information we could report is the frequency with which every variable condition appeared in the data. However, if we had 20 speakers aged 30 years and 20 speakers aged 32 years, we could say, more than on average, the participants are 31 years old.

A detailed description of the suitability of the different types of variables for specific statistical tests is beyond the scope of this chapter, but it is important to know that statistical tests can be classified into two main categories:

– parametric tests;

– non-parametric tests.

Parametric tests can be used when the linguistic variable under consideration is measured on a scale, and the explanatory variable is measured following a nominal or an ordinal criterion. These tests include T-tests, used when the explanatory variable has two categories, and ANOVA – Analysis of Variance – used when the explanatory variable has more than two categories. For example, this type of test can be used in studies seeking to find out whether advanced learners produce significantly more discourse connectives than intermediate learners, or whether speakers produce more fallacious arguments when speaking at political party meetings or at electoral campaigns. Such tests will not be discussed in this chapter since they are rarely used in corpus linguistics, but you can refer to Brezina (2018) for a presentation on them.

When the linguistic variable is measured on a nominal or ordinal scale, it is necessary to turn to non-parametric tests, such as the chi-square test (or chi^2). For example, in order to determine whether type of speech acts varies between two text genres, it is necessary to carry out a chi-square test. It is this test that we will present and illustrate in the rest of this chapter.

Table 8.5 shows examples of the types of variables introduced in this section.

	Linguistic	**Explanatory**
Nominal	Pronoun category (relative, interrogative, personal, etc.)	Gender (male, female)
Ordinal	Groups of children classified according to the mean length of utterances (<2, 2–3, >3)	Language proficiency level (beginner, intermediate, advanced)
Scalar	Relative frequency of passive sentences in a corpus	Age of speakers (measured along a *continuum*)

Table 8.5. *Examples of relevant variables in corpus linguistics and their types*

8.6. Measuring the differences between categories

In this section, we will focus on how to report frequency differences between categories in order to determine whether these differences are significant, and then to establish a link between two nominal or ordinal variables, for example, whether the value of a nominal or an ordinal variable depends on another nominal or ordinal variable.

Let us imagine that we are conducting research on the use of regionalisms in French. For example, it is widely known that some French-speaking regions such as Switzerland and Belgium use different words, from those used in France, for designating numbers such as 70 and 90, in this case *septante* and *nonante*. The situation seems a little more complicated in the case of 80, which is expressed as *huitante* in some Swiss cantons and *quatre-vingts* in others. According to a recent linguistic atlas of regional French (Avanzi 2017), the Swiss cantons using the word *huitante* are Vaud, Freiburg and Valais, whereas the cantons of Geneva, Jura and Neuchâtel tend to use the word *quatre-vingts*.

Different questions can be studied concerning this situation. First, we should check with corpus data whether the use of *huitante* varies from one group of cantons to another. To do this, we have to look for the number of occurrences of the word *huitante* in each of these six cantons by means of the OFROM corpus, which compiles the French spoken in Switzerland (see Chapter 5). Then, we have to group the results in a table such as Table 8.6, which follows the group classification used in the linguistic atlas – *huitante* (Vaud, Valais, Freiburg) versus *quatre-vingts* (Neuchâtel, Jura, Genève) – and transform these results, in order to make them comparable by dividing by the total number of words produced per canton (as indicated in the metadata of the corpus) and multiplying by a base of normalization.

	***Huitante* cantons**	***Quatre-vingts* cantons**	**Total**
huitante	32.73 (40)	6.92 (10)	**39.65 (50)**

Table 8.6. *Frequency of the word "huitante" every 100,000 words per group of cantons (number of occurrences in brackets)*

At a second stage, to find out whether the number of occurrences of *huitante* is different from one group of cantons to another, it is necessary to compare the observed values with the expected ones. The values observed are those found in the corpus. The expected values are those that should be found if the null hypothesis was correct and there was no difference in the use of the word *huitante* between the two groups of cantons. Here, the frequency of *huitante* is equal to 39.65 in the entire corpus. Since there are two groups of cantons, the expected value for each group should be 19.83 (3.65/2), which is not at all what we observed in our data.

In order to test whether this difference is significant, we use a statistical test called the chi-square goodness-of-fit test, represented by the symbol χ^2 due to the notation denoted by the Greek letter that gave it its name.

The χ^2 is actually a value calculated according to the following formula, where Σ denotes the sum of all categories considered:

$$\chi^2 = \Sigma \frac{(\text{observed value} - \text{expected value})^2}{\text{expected value}}$$

The value of χ^2 in our example therefore corresponds to:

$$\frac{(32.73-19.83)^2}{19.83} + \frac{(6.92-19.83)^2}{19.83} = 16.8$$

This value makes it possible to know the probability of such results actually being observed, if the occurrences were randomly distributed along the variable categories (which would then represent the null hypothesis). The higher the value of χ^2, the more it suggests that the distribution observed moves away from the expected one if the distribution is uniform. The value obtained should be compared with a critical value of χ^2, which in this case depends not only on the field (which establishes the significance threshold) but also on the number of degrees of freedom (often indicated as "df"). The number of degrees of freedom is calculated by subtracting 1 from the number of categories in the explanatory variable. For a variable with two categories, as we have here, the χ^2 test is significant with a threshold placed at 0.05 if the value of χ^2 is greater than 3.84. Thanks to the corpus data considered, we can deduce that (at 5%) the use of *huitante* is almost certainly associated with a group of cantons, which corroborates the assertion made on the atlas we quoted.

To apply the χ^2 test, we have to calculate the χ^2 coefficient value, which can be done manually, as we did here, or by using statistical software. The value can also be calculated online using on the VassarStats site, by choosing the option "Chi square goodness of fit" accessible in the "Frequency data" tab. To do this, in this example, the frequencies observed (Table 8.6) must be entered as inputs in to the calculator, as well as the expected frequencies (for the two groups of cantons). In order to report the results of the test, we usually specify the number of degrees of freedom of the χ^2 in brackets, indicating the coefficient value, as well as the degree of significance (p) associated with this value, as in the following sentence: "The distribution of the word *huitante* is not uniform between the two groups of cantons ($\chi^2(1) = 16.8$, $p < 0.001$). This word is used significantly more in the group of cantons of Freiburg, Vaud and Valais".

The χ^2 can also be calculated for variables with more than two categories. For example, we might want to determine whether the use of *huitante* varies between the three cantons that are supposed to use this word, namely Vaud, Valais and Freiburg. This time, we should type the observed values (Table 8.7) and the expected values (32.73 / 3 = 10.91) into the calculator.

	Vaud	**Valais**	**Freiburg**	**Total**
huitante	16.44 (21)	8.31 (10)	7.98 (9)	**32.73 (40)**

Table 8.7. *Relative frequency of "huitante" every 100,000 words for the three cantons where this word is used*

The result of the test for these values is as follows: $\chi^2(2) = 4.21$, $p = 0.12$. This time, the degrees of freedom are 2 because there are three categories for the explanatory variable. Unlike the previous example, the value of p is not smaller than 0.05, which indicates that the result is not statistically significant. This result can be reported as follows: "The three cantons where the word *huitante* is used (Vaud, Valais and Freiburg) do not differ significantly in the use they make of this word ($\chi^2(2) = 4.21$, $p = 0.12$)".

However, through the observation of the data described above, we cannot yet know whether the cantons vary in their frequency of use of the two regional words *huitante* and *quatre-vingts*. To find out, we should draw a comparison between the use of *huitante* and *quatre-vingts* in the two groups of cantons. To do this, we have to count the occurrences of *huitante* and *quatre-vingts* and relate these to the two cantonal categories in a crosstab.

To build a crosstab, we only have to place one of the variables in columns (often the explanatory variable) and the other variable in rows (the linguistic variable). For the question we are focusing on here, Table 8.8 shows how to organize data by placing the two categories of the explanatory variable (the groups of cantons described in the linguistic atlas) in a column and the two categories of the linguistic variable (the expression for denoting 80) in a row.

	Huitante cantons	*Quatre-vingts* cantons	Total
Occurrences of *huitante*	40	10	**50**
Occurrences of *quatre-vingts*	36	46	**82**
Total	76	56	**132**

Table 8.8. *Number of occurrences of the two words used for expressing 80, by group of cantons*

In order to normalize these figures and to summarize data, it is possible to transform the different categories into percentages either by columns, by rows or even in relation to the global total of occurrences. The type of percentages used depends on the aspect of the table we wish to illustrate. For example, in order to describe the differences between cantons, this percentage must be calculated on the basis of the columns. Such a transformation illustrates the fact that in the cantons where people are supposed to use the word *quatre-vingts* rather than *huitante*, according to the linguistic atlas, we only find a 22% use of *huitante* compared to the combined uses of *huitante* and *quatre-vingts* (10/46), whereas in the cantons supposed to use *huitante*, this proportion rises to 53% (40/76). These proportions thus seem to confirm the information from the linguistic atlas.

A representation of percentages in relation to the grand total implies a total transformation of the data, in which each cell is divided by the grand total (132). The data thus transformed are not intrinsically more informative than those in Table 8.8, but the transformation into percentages makes it possible for proportions to be better represented than raw data. We should be aware, however, that if this type of format is chosen, the raw data that gave rise to the percentages must always be visible, for example, by indicating them within brackets after the percentages, as in Table 8.9.

	Huitante cantons	*Quatre-vingts* cantons	Total
huitante	30% (40)	8% (10)	**38% (50)**
quatre-vingts	27% (36)	35% (46)	**62% (82)**
Total	**57% (76)**	**43% (56)**	**100% (132)**

Table 8.9. *Occurrences of regional words by groups of cantons expressed in percentages*

We have just presented a very simple crosstab in which each variable has only two categories. Such a table could just as easily be produced for variables with more categories, for instance, if we had processed the values of each canton separately, rather than synthesizing them into two groups, as in Table 8.10. However, a crosstab has, by definition, only two dimensions and can therefore represent only two variables in rows and columns, respectively.

	Vaud	**Valais**	**Freiburg**	**Neuchâtel**	**Jura**	**Geneva**	**Total**
huitante	21	10	9	6	2	2	**50**
quatre-vingts	15	14	7	35	8	3	**82**
Total	**36**	**24**	**16**	**41**	**10**	**5**	**132**

Table 8.10. *Number of regional word occurrences per canton*

Once the data has been inserted into a crosstab, the next step is to determine whether the distribution of one variable into categories depends on the other variable. The test that answers this question is very similar to the one presented above, except that this time it aims to test the degree of independence between two variables. For this reason, it is called the χ^2 test of independence. It is based on the same formula as above. The difference lies in the way of calculating the expected value for each cell of the table and the number of degrees of freedom. Since this test compares differences in proportions between two categories, each of them drawn from the same portion of the corpus, this time, it is the raw number of occurrences that should be used, rather than transformed values.

The expected value for a cell in the table is calculated by multiplying the total of its row and that of its column, and then dividing the result by the grand total. For example, the expected value for the first cell in Table 8.9, which has an observed value of 40, is obtained by multiplying 50 (sum of the values in the row) by 76 (sum of the values in the column), and dividing by 132 (grand total); that is, a value of 2,879. The different frequencies expected for this table are summarized in Table 8.11.

	Huitante **cantons**	***Quatre-vingts*** **cantons**	**Total**
huitante	40 (28.79)	10 (21.21)	**50**
quatre-vingts	36 (47.21)	46 (34.79)	**82**
Total	76	56	**132**

Table 8.11. *Observed (and expected) frequencies of the two words used for denoting 80, per group of cantons*

The value of the χ^2 coefficient corresponding to this data is:

$$\frac{(40-28.79)^2}{28.79}+\frac{(10-21.21)^2}{21.21}+\frac{(36-47.21)^2}{47.21}+\frac{(46-34.79)^2}{34.79}=16.57$$

As in the χ^2 goodness-of-fit test, this value should be compared to a critical χ^2 for a threshold conventionally fixed at 0.05, and the appropriate number of degrees of freedom. In the case of the crosstab, the number of degrees of freedom is calculated according to the (L-1) (C-1) formula, where L corresponds to the number of rows of the table and C corresponds to the number of columns.

In this example, the number of degrees of freedom is equal to (2-1) (2-1), that is, 1. The critical value is 3.84 once again, so the result of the test is significant. This indicates that the distribution of the words *quatre-vingts* and *huitante* between the two groups of cantons is not random, but clearly reflects a real difference in linguistic practices between them.

The value of the χ^2 test of independence can also be calculated using statistical software, or directly online using the same site mentioned previously. This time, on VassarStats, choose the option “Chi square, Cramer’s V and Lambda” which is accessible on the “Frequency data” tab, enter the values observed in the table and choose the option “Calculate”. Here is what the VassarStats site retrieves for the values in Table 8.8.

Chi-Square	df	P
15.12	1	0.0001

Cramer's V = 0.3543

Note that for df=1 the chi-square value reported is the Yates chi-square, corrected for continuity.
The Pearson chi-square, uncorrected for continuity, is 16.57
P = <.0001

Figure 8.2. *Chi-square test result, as displayed in VassarStats*

We note that the value of χ^2 indicated on the left is not the one we found because this value has been slightly rectified, as indicated on the right (Yates correction). However, the latter is not generally used in corpus linguistics, so it is the value that we have found and that is indicated in the dialog box on the right (without correction) that we will use. Finally, the test result indicates that the probability *p*, with which we can confirm the hypothesis given the value of χ^2 obtained, here, is smaller than 0.0001. As this value is clearly smaller than 0.05, this confirms that the test result is significant at 5% and even at 0.01%. The result of this test should be reported as follows: "We carried out the χ^2 test of independence in order to determine whether the two groups of cantons varied in their use of regional words *quatre-vingts* and *huitante*. The results indicate that the use of these words varies significantly between cantons. $\chi^2(1) = 16.57$, $p = 0.001$". Note also that, by convention, values of p smaller than 0.001 are not reported exactly, but summarized by the indication $p < 0.001$.

The indication "Cramer's V" is a measurement of the size of the observed effect. This is an indication of the degree of relation between two variables. The result of the χ^2 test provides only one answer to the existence of a link between two variables, but not to the importance of this association. For a crosstab with one degree of freedom (2x2), we consider that the effect is small when the value of Cramer's V reaches 0.1, moderate when the value reaches 0.3 and large when the value is 0.5 or more (Cohen 1988). For a table with two degrees of freedom, the effect is small from 0.07, moderate from 0.21 and large from 0.35. At three degrees of freedom, the threshold is lowered further. We describe this as a small effect from 0.06 onwards, a moderate effect from 0.17 and a large effect from 0.29. In the case of our test, we observe a moderate effect (V = 0.35). The link between the two variables examined is of moderate importance.

Let us then imagine that we wish to know whether the use of the words *huitante* and *quatre-vingts* varies specifically between the three cantons that use the word *huitante* significantly less, namely Neuchâtel, Jura and Geneva. In order to find this out, if we enter the figures in Table 8.10 for these three cantons in an χ^2 calculator, the latter will return an error message, or at least a warning. In fact, the χ^2 test offers reliable results only when at least 80% of the expected values are greater than or equal to 5. However, this constraint is not fulfilled here because three cells have expected values that are smaller than 5 (3.79 for *huitante* in Jura, 1.89 for *huitante* in Geneva and 3.11 for *quatre-vingts* in Geneva), which represents 50% of the cells (3 out of 6). In this type of case, another test must be carried out which does not pose the same constraint.

The suitable test is called Fisher's exact test, which can be calculated using the same online statistical toolboxes as the χ^2 test. This test offers only a probability and not a statistical value because it is an exact test. In this case, the value of Fisher's exact test is $p = 0.29$, which indicates that the distribution between these two words does not vary significantly between these three cantons. For this test, the sentence reporting this result could be as follows: "The speakers of the cantons of Neuchâtel, Jura and Geneva do not vary significantly in their use of the words *huitante* and *quatre-vingts* (two-sided Fisher's exact test, $p = 0.29$)".

Finally, note that the chi-square test of independence also makes it possible to know whether all the cantons differ from one another or not. To carry out this test, let us set aside the values of the canton of Geneva, which is under-represented in the corpus, with only 75,598 words (against more than 100,000 in all the other cantons). This poses problems for carrying out an χ^2 test, as we have just seen. The result of the χ^2 test of independence for the five cantons of Vaud, Valais, Freiburg, Neuchâtel and Jura is as follows: $\chi^2(4) = 19.63, p = 0.001$. We can see that the difference in the distribution of the two words between the cantons is significant.

However, this result is not entirely informative. Indeed, the only thing that this test lets us conclude is that the distribution between cantons differs significantly in some way, or, in other words, that it is unlikely to observe such a biased distribution under the assumption that *huitante* and *quatre-vingts* are used uniformly across the five cantons. Nevertheless, the test does not tell us whether all the cantons differ from one another or if only some of them are different from the others, nor whether this difference is linked to the use of the two words or only to one of them.

To determine where a significant difference stems from when an χ^2 test comprises variables from more than two categories, we have to observe standardized residual values of the test. A standardized residual is a ratio calculated from the observed frequency and the expected frequency for each category, according to the following formula:

$$\text{standardized residual} = \frac{\text{observed frequency} - \text{expected frequency}}{\sqrt{\text{expected frequency}}}$$

The more the residual moves away from 0, the more it means that the category contributes to the significant test result. According to a rule of thumb, we consider that any value of a standardized residual greater than +2 or smaller than -2 shows a significant difference compared to the mean, because this figure means that the value obtained deviates by more than two standard deviations from the mean. A positive value reflects an overuse compared to the other categories, whereas a negative value reflects an underuse. In VassarStats, standardized residuals are calculated automatically and reported below the test results table. In the case of the test that we have to perform, the standardized residuals obtained are shown in Table 8.12.

	Vaud	Valais	Freiburg	Neuchâtel	Jura
huitante	+ 2	+ 0.31	+ 1.2	+ 2.41	+ 0.92
quatre-vingts	+ 1.56	+ 0.24	+ 0.94	+ 1.88	+ 0.71

Table 8.12. *Standardized residuals for the χ^2 test corresponding to Table 8.10, without the canton of Geneva*

By following the rule of thumb presented above, we see that the significant difference in the test comes from a much more frequent use of the word *huitante* in the canton of Vaud, and a much less frequent use in the canton of Neuchâtel, compared to the other cantons. The other cantons do not differ significantly. The distribution of the word *quatre-vingts* is never significantly different between cantons.

8.7. Conclusion

In this chapter, we have discussed some simple procedures that help to report and analyze the numerical values obtained in a corpus study. We have made an important distinction between descriptive statistics (which relate only to the sample tested) and inferential statistics (which aim to generalize the results to an entire population). We have also described the different forms that variables in a corpus may adopt, and highlighted the fact that statistical tests must be adapted to the nature of the variables. In terms of descriptive statistics, we started by presenting the advantages and disadvantages of the mean and the median as a way of synthesizing data and concluded that the use of the mean also requires indicating the dispersion of the values around it (notion of standard deviation). We then tackled the question of data normalization, which makes it possible to compare values from different corpora, transforming them into relative frequencies through the use of bases of normalization. We have emphasized that frequency is not the only indicator of the prevalence of a word in a corpus and that its dispersion across the different portions of the corpus is also very important. We therefore introduced a simple and a more sophisticated way (deviation of proportions) to calculate this dispersion. We also showed how to measure the lexical richness of a corpus using the type/token ratio, and presented the limitations of such a measurement. To conclude, we gave two variants of the chi-square test, an inferential statistics test which makes it possible to determine whether the differences observed between the distributions of several categories are significant.

In the case of the chi-square test of independence, we have shown how to represent data using a crosstab. We introduced the notion of standardized residual, which makes it possible to determine where the significant result of the test comes from, and introduced Fisher's exact test as an alternative to the chi-square, when the conditions posed by the latter are not met.

8.8. Revision questions and answer key

8.8.1. *Questions*

1) The tables below show the number of occurrences of the (lemmatized) words *bateau* and *je*, as well as their English equivalent *boat* and *I*, in the bilingual corpus of children's literature *ParCoGLiJe* (see Chapter 5). This corpus includes four original works in French and their translation into English, as well as four original works in English and their translation into French. Raw data (number of occurrences of the words *bateau*, *je*, *boat*, *I*) drawn from these works are presented in the four tables below, as well as number of word types and word occurrences. Normalize the data so as to be able to compare the frequency of these words throughout the texts in each table, and then between the sub-corpora (original texts vs. translated texts, French texts vs. English texts).

Original texts in French

	bateau	***je***	**Word types**	**Word occurrences**
Vingt mille lieues sous les mers	55	2,863	14,629	148,563
Mémoires d'un âne	8	1,848	5,520	56,113
Les trois mousquetaires	4	3,606	13,926	238,530
Lettres de mon moulin	10	585	7,388	49,592

Texts translated into French

	bateau	*je*	**Word types**	**Word occurrences**
Le jardin secret	2	1,274	6,836	80,737
L'île aux trésors	14	1,734	8,597	72,706
Le livre de la jungle	0	609	6,753	57,619
Oliver Twist	3	1,720	11,676	169,016

Original texts in English

	boat	*I*	**Word types**	**Word occurrences**
The Secret Garden	1	1,426	4,786	83,483
Treasure Island	61	1,978	5,845	70,431
The Jungle Book	3	650	4,575	52,207
Oliver Twist	2	1,868	10,186	161,327

French texts translated into English

	boat	*I*	**Word types**	**Word occurrences**
Twenty Thousand Leagues Under the Sea	61	2,944	11,865	144,435
Cadichon's Life Story	5	1,330	4,067	42,186
The Three Musketeers	24	3,848	10,217	231,925
Letters From My Mill	22	646	6,594	48,404

2) On the basis of the four above-mentioned sub-corpora, what hypotheses could be formulated concerning:

– the difference in distribution between the words *bateau* and *je* in French;

– the difference between original texts and translated texts;

– the difference between French and English.

What are the corresponding null hypotheses?

What are the types of variables corresponding to:

– the number of occurrences of *bateau*, *je, boat, I*;

– the type of text: translated or original;

– the novel: *Le jardin secret, L'île au trésor, Oliver Twist, Le livre de la jungle*;

– the number of word types in each corpus;

– the language: French or English.

3) Calculate the type/token ratio for each text. In which case is it methodologically correct to apply this measure?

4) Calculate lexical dispersion for the words *boat, I, bateau* and *je* in each of the four sub-corpora using the deviation of proportions method. What can you notice?

5) Determine whether the distribution of different translations of *néanmoins* and *toutefois* found by Dupont and Zufferey (2017) (numbers shown in Table 8.3) differs when we carry out a chi-square test. Consider *nonetheless* and *nevertheless* as two lexical variants of the same connective and group them together. If this distribution is significantly different, where do these differences come from? How big is the effect?

8.8.2. *Answer key*

1) First, let us recall that in order to normalize data, we have to divide the number of occurrences of the word we are interested in by the total number of words in the corpus, and then we multiply this result by a base of normalization. Given the size of the different texts, we chose a base of 10,000 words.

Relative frequency every 10,000 words of lemmas bateau *and* je *in French texts (originals and translations)*

	bateau	*je*
Vingt mille lieues sous les mers	3.70	192.71
Mémoires d'un âne	1.25	246.82
Les trois mousquetaires	0.17	151.18
Lettres de mon moulin	2.02	117.97
Le jardin secret	0.25	157.80
L'île aux trésors	1.93	238.49
Le livre de la jungle	0	79.49
Oliver Twist	0.18	39.35

Relative frequency every 10,000 words of lemmas boat *and* I *in English texts (originals and translations)*

	boat	***I***
Twenty Thousand Leagues Under the Sea	4.22	203.83
Cadichon's Life Story	1.19	315.27
The Three Musketeers	1.03	165.92
Letters From My Mill	4.55	133.67
The Secret Garden	0.12	170.81
Treasure Island	8.66	280.84
The Jungle Book	0.57	124.50
Oliver Twist	0.12	115.79

The normalized data enables us to observe that the pronouns *je* and *I* have higher frequency than the nouns *bateau* and *boat*. This result is widely expected, insofar as closed-class words usually have higher frequencies than open-class words, as is the case of nouns. We can also observe that the use of *je* and *I* is highly variable between novels. This gap reflects the different narrative perspectives used. First-person novels make much more use of it than third-person novels. There are also great differences between certain texts and their translation, especially in the case of *Mémoires d'un âne*, where *I* appears three times less frequently in the English translation than in the French. Finally, it is interesting to note that the frequency of the words *bateau* and *boat* also varies between the original texts and their translation. This seems to indicate that the semantic fields of these words are not the same at all.

For example, in French, the word *sous-marin* appears regularly in the novel *Vingt mille lieues sous les mers* and this word is translated as *underwater boat* in English. So, in English, the word *boat* is used for qualifying this means of transport, but this is not the case in French. A more in-depth analysis would reveal other interesting differences in the use of these words between languages, as we argued in Chapter 4.

2) Here are some examples of hypotheses, but many others could be possible:

H1: the word *je* has a higher frequency than *bateau* since it is a function word;

H0: there is no difference in frequency between the words *je* and *bateau*;

H1: translated texts are lexically less diversified than original texts due to the simplification universal;

H0: translated texts and original texts have the same level of lexical diversity;

H1: French texts are longer (have more word occurrences) than English texts;

H0: texts in French and English have a similar length.

3) a) The number of occurrences of the words *bateau* and *je* is a scalar-type linguistic variable.

b) The type of text, either translated or original, is a nominal-type explanatory variable.

c) The novel is a nominal-type explanatory variable.

d) The number of word types in each corpus is a scalar-type linguistic variable.

e) Language is a nominal-type explanatory variable.

	type/token ratio
Vingt mille lieues sous les mers	0.10
Mémoires d'un âne	0.10
Les trois mousquetaires	0.06
Lettres de mon moulin	0.15
Le jardin secret	0.08
L'île aux trésors	0.12
Le livre de la jungle	0.12
Oliver Twist	0.07
Twenty Thousand Leagues Under the Sea	0.08
Cadichon's Life Story	0.10
The Three Musketeers	0.04
Letters From My Mill	0.14
The Secret Garden	0.06
Treasure Island	0.08
The Jungle Book	0.09
Oliver Twist	0.06

The type/token ratio can only be used for comparing texts of similar length. For example, it is suitable for comparing *Lettres de mon moulin* and *Mémoires d'un âne*, but not *Lettres de mon moulin* and *Les trois mousquetaires*, since the latter is more than four times longer. In fact, this ratio decreases as the texts lengthen, which is true for the case of these two novels. The type/token ratio can also be used for comparing texts and their translations, only in cases of similar length.

Proportional differences for the words *bateau* and *boat*:

	bateau	***boat***
Original	0.44	0.73
Translated	0.61	0.33

We can observe that the word *bateau* in the original texts of the four novels in French and in their English translations is used more homogeneously (0.44 and 0.33 respectively) for the word *boat* than in the original English texts and their French translations (0.73 and 0.61 respectively). This difference is partly due to the fact that the word *bateau* is found in the four novels written in French, whereas it is only found in three of the four novels written in English.

Proportional differences for the words *je* and *I*:

	je	***I***
Original	0.11	0.18
Translated	0.42	0.09

We can see that the dispersion of pronouns *je* and *I* is much more homogeneous than that of the noun *bateau*, as we could expect, given the high frequency of this word. The great homogeneity of this distribution in the texts written in French can also be found in the translations of these texts into English. However, while the texts written in English make homogeneous use of *I*, their translation into French presents a much less homogeneous use of *je*.

5) The application of a chi-square test of independence indicates that the distribution of the different translations is not homogeneous between the two connectives ($\chi^2(4) = 14.06$, $p < 0.01$). The standardized residuals also indicate that this difference is due to an overuse of *nevertheless/nonetheless*

when it comes to translating *néanmoins* (+ 2.41). The other differences are not significant. The effect size is moderate (Cramer's V = 0.22).

8.9. Further reading

Brezina (2018) is one of the best introductory works on the use of statistics for corpus data. This book is also accompanied by a website that offers access to statistical tools for carrying out most of the tests presented here. Gries (2013) is also an introduction to the use of statistics for corpus linguistics, in connection with the use of R software. Field (2017) is an excellent introduction to statistics in general and to the use of SPSS, another kind of software for performing statistical tests.

Conclusion

The Stages for Carrying Out a Corpus Study

Throughout this work, we have presented the different facets of corpus linguistics, both from the point of view of the theoretical questions to which this discipline provides answers and of its methodological foundations. By way of conclusion, we would like to offer a list of ordered stages, making it possible to implement the concepts discussed in this book step by step, and to carry out a corpus study.

C.1. Stage 0: wanting to know more

Before starting any project, something that is important is the desire to learn more about it. It is actually this initial curiosity that gives birth to the best research ideas. Before starting to work on a project, it is essential to try to find a question you are interested in, or at least which arouses your curiosity. Often, this first idea is intuitive and rather vague.

For example, some are fascinated by the question of the differences between men and women and how they are reflected in language. Others are enthusiastic about children and everything related to them. Still others like politics, history or sport. The great advantage of linguistics is that the study of language has interfaces with very many disciplines, and that it is possible to find study subjects in very varied fields. Take the time to question your interests and listen to your intuitions, even if the latter do not (yet) look like a research question that can be studied empirically.

C.2. Stage 1: identify relevant literature

Once you have found the subject you are interested in, the first real research step will be to find relevant documentation. Indeed, corpus-based studies aim to test empirical hypotheses that are often formulated from the results of other studies, either more theoretical or empirical in nature, which have already been published in the scientific literature.

Thus, the first step in a corpus study is to identify relevant sources that have so far explored the research subject under consideration. Most of these sources can be found in scientific journals. For example, journals such as *Corpus Linguistics and Linguistic Theory*, *International Journal of Corpus Linguistics* and *Corpora* are all three specialized in the publication of corpus studies, whereas the *Journal of Language Resources and Evaluation* publishes articles on methodological aspects related to the compiling and processing of corpus data. In addition, many journals specializing in different areas of linguistics, such as *Journal of Pragmatics*, *Journal of Sociolinguistics* and *Journal of French Language Studies*, regularly publish corpus studies. We have offered many examples of such studies in Chapters 2–4. The names of the journals or books in which these studies have been published can easily be found by consulting the reference list at the end of this book.

The journal sites mentioned above will help you to search for articles using keywords. Another way to identify relevant literature is to use the Google Scholar search engine, which indexes most of the available scientific articles. Other databases that are available for a fee, such as the Scopus database, make it possible to search for articles from many different journals using a single query.

In scientific journals, in most cases, access to articles is not free, although the alternative model called Open Access is gaining popularity. University libraries generally offer access to many online journals and should be checked for availability via the university computer network. When an article comes from a journal that is not freely available, it can still be accessible in certain cases. In fact, more and more researchers are making a version of their articles available to the public on sites such as ResearchGate and Academia, or on the site of their institution or on personal web pages. It is therefore useful to look for the article title directly in a search engine to find out whether such a version is available online.

C.3. Stage 2: formulating research hypotheses

Once the relevant studies have been identified in the scientific literature, the second stage in a corpus study is to formulate specific hypotheses to study the research question under consideration. These hypotheses emerge from the results of previous studies, and also aim to supplement them, for example, by testing new variables.

Let us recall that the hypotheses must be empirically testable (see Chapter 8, section 8.4). In other words, they should clearly identify variables and suggest relations between them. For every research hypothesis, it is also important to formulate the corresponding null hypothesis, since it is the latter that will be evaluated by means of statistical tests.

C.4. Stage 3: operationalizing your hypotheses and choosing data

The operationalization phase of research hypotheses is crucial for the success of an empirical study. This phase consists of determining how the variables will be measured. For example, if the question to be investigated is the assertion that women talk more about their emotions than men, the operationalization of this question will require building or obtaining a lexicon of words related to emotions that can be searched in a corpus, chosen to make it possible to determine whether women actually use these words more than men.

As from the operationalization phase, it is also important to choose the corpus on which the study will be carried out. If the study makes use of existing corpora, the possibility of investigating a certain question or not, or the manner in which it can be operationalized, depends on the characteristics of the corpus.

For example, to be able to compare the language of women and that of men, this variable must be clearly identifiable in the metadata of a corpus. Likewise, the stylistic genre of the corpus should be compatible with the question under investigation. People are more likely to talk about their emotions in spontaneous conversations or when they relate a memory than when they are giving a scientific presentation, for example. The operationalization phase must therefore lead to the decision to use an existing resource, for example among those described in Chapter 5 or, to create a new corpus, according to the principles introduced in Chapter 6.

C.5. Stage 4: extracting and annotating corpus data

Once the corpus has been identified or created, the next step is to determine the best strategy for retrieving the relevant data. In order to automatically search for elements in a corpus, a surface feature such as a word or list of words should be associated with it. In the case of the research question mentioned above, it would be the vocabulary of emotions.

In many cases, the automatic extraction includes data which are irrelevant for analysis and which require manual sorting. For example, not all the uses of the verb *like* correspond to the expression of an emotion, because in certain cases this verb is used with a modal value (*I would like you to eat your ice cream*). Data extraction can also be based on a prior automatic analysis of the chosen corpus, such as lemmatization or part-of-speech tagging (see Chapter 7, section 7.4).

For many questions, the raw data retrieved from a corpus will not be sufficient. It will therefore be necessary to annotate data before obtaining answers to these questions. The annotation process requires the prior identification of clear categories (see Chapter 7, section 7.3). A first pilot annotation phase should be used for testing these categories, as well as for identifying problematic cases and leading to the preparation of a more or less detailed annotation manual. In order to be reliable, an annotation should ideally be carried out by several annotators independently, and their measured agreement should be placed above a certain threshold (see Chapter 7, section 7.5).

C.6. Stage 5: analyzing data

The data analysis stage must begin with any transformations in order to make the data comparable across corpora. This stage notably involves the transformation of raw figures corresponding to the number of occurrences observed in the corpus into figures reporting relative frequencies, following a base of normalization. Then, different descriptive statistics should be carried out. In order to be able to draw quantitative conclusions, it is necessary to carry out inferential statistical tests, which will indicate whether the results obtained on the corpus sample can be generalized to the entire population.

We should recall that the possibility of using different tests depends on the way in which the hypotheses have been operationalized and, more specifically, on the types of variables involved (see section 8.4).

C.7. Stage 6: presenting your study in a report or an article

Corpus studies examine empirical data in order to draw quantitative conclusions, which are to be interpreted in the light of the hypotheses formulated in stage 2. The reports or articles presenting this type of results generally follow a very precise structure. In the introduction to the document, the research question is briefly introduced and contextualized. The second section contains a review of relevant previous studies (also called the state of the art), which served as a theoretical basis for the study, or which are inspired on other corpus studies that the current research project will supplement or sometimes call into question. The third section contains a presentation of the hypotheses resulting from it and which will be the subject of the empirical study. The rest of the document aims to present the results of the study. The detailed presentation of the study begins by describing the corpus used and explaining how data was retrieved from it, as well as their annotation. Then, the results are presented in detail, as well as the statistical analyses that were carried out, with the results of the statistical tests reported in accordance with a very precise format (see section 8.6). These results should finally be discussed in a critical manner, indicating, in particular, the extent to which the initial hypothesis is verified. A brief conclusion can summarize the main results obtained, or even provide perspectives for further studies. This order of presentation can be summarized in a diagram as follows:

1) introduction;

2) state of the art;

3) hypotheses;

4) corpus study:

   a) data,

   b) procedure: extraction and annotation,

c) results;

5) discussion of the study;

6) conclusion.

C.8. Conclusion

Once all these steps have been completed, you will have the satisfaction of having contributed to research by providing the scientific community with new empirical data, which in turn can serve as a starting point for other research. Your results could be replicated and reassessed by other researchers. Very often, the results of empirical studies also serve to modify and improve existing theories, and thus contribute to make linguistics a scientific study of language. This is one of the key objectives of the empirical approach that we presented in Chapter 1: to provide a scientific perspective on language, using a rigorous methodological approach based on the quantitative analysis of linguistic data.

References

Aarts, B. (2001). Corpus linguistics, Chomsky and fuzzy tree fragments. In *Corpus Linguistics and Linguistic Theory*, Mair, C. and Hundt, M. (eds). Rodopi, Amsterdam, 5–13.

Abeillé, A., Clément, L., and Toussenel, F. (2003). Building a treebank for French. In *Building and Using Parsed Corpora*, Abeillé, A. (ed.). Springer, Dordrecht, 165–187.

Adolphs, S. and Knight, D. (2010). Building a spoken corpus: What are the basics? In *The Routledge Handbook of Corpus Linguistics*, O'Keeffe, A. and McCarthy, M. (eds). Routledge, Abingdon-on-Thames, 38–52.

Afantenos, S., Asher, N., Benamara, F., Bras, M., Fabre, C., Ho-Dac, L.-M., Le Draoulec, A., Muller, P., Péry-Woodley, M.-P., Prévot, L., Rebeyrolle, J., Tanguy, L., Vergez-Couret, M., and Vieu, L. (2012). An empirical resource for discovering cognitive principles of discourse organization: The ANNODIS corpus. *Proceedings of the Eighth International Conference on Language Resources and Evaluation (LREC'12).* European Language Resources Association (ELRA), Istanbul, 2727–2734.

Aijmer, K. (2008). Comparable and parallel corpora. In *Corpus Linguistics: An International Handbook*, Lüdeling, A. and Kytö, M. (eds). Walter de Gruyter, Berlin, 275–292.

Aijmer, K. (2009). *Corpora and Language Teaching*. John Benjamins, Amsterdam.

Altenberg, B. (1999). Adverbial connectors in English and Swedish: Semantic and lexical correspondances. In *Out of Corpora. Studies in Honour of Stig Johansson*, Hasselgård, H. and Oksefjell, S. (eds). Rodopi, Amsterdam, 249–268.

Artstein, R. (2017). Inter-annotator agremment. In *Handbook of Linguistic Annotation*, Ide, N. and Pustejovsky, J. (eds). Springer, Dordrecht, 297–313.

Asher, N. and Lascarides, A. (2003). *Logics of Conversation*. Cambridge University Press, Cambridge.

Aston, G. (2001). Learning with corpora: An overview. In *Learning with Corpora*, Aston, G. (ed.). Athelstan, Houston, 4–45.

Aston, G. and Burnard, L. (1998). *The BNC Handbook*. Edinburgh University Press, Edinburgh.

Atkins, B. and Rundell, M. (2008). *The Oxford Guide to Practical Lexicography*. Oxford University Press, Oxford.

Avanzi, M. (2017). *Atlas du français de nos régions*. Armand Colin, Paris.

Avanzi, M., Béguelin, M.-J., and Diémoz, F. (2017). Présentation du corpus OFROM – corpus oral de français de Suisse romande Université de Neuchâtel, Neuchâtel [Online]. Available: http://www.unine.ch/ofrom.

Baayen, R.H. (2008). Corpus linguistics in morphology: Morphological productivity. In *Corpus Linguistics. Handbook of Linguistics and Communication Science*, Lüdeling, A. and Kytö, M. (eds). de Gruyter Mouton, Berlin, 899–919.

Baayen, R.H., van Halteren, H., and Tweedie, F. (1996). Outside the cave of shadows: Using syntactic annotation to enhance authorship attribution. *Digital Scholarship in the Humanities*, 11(3), 121–132.

Baker, M. (1993). Corpus linguistic and translation studies: Implications and applications. In *Text and Technology*, Baker, M., Francis, G., and Tognini-Bonelli, E. (eds). John Benjamins, Amsterdam, 233–250.

Baker, P. and Vessey, R. (2018). A corpus-driven comparison of English and French Islamist extremist texts. *International Journal of Corpus Linguistics*, 23(3), 255–278.

Belica, C. (1996). Analysis of temporal change in corpora. *International Journal of Corpus Lingusitics*, 1(1), 61–74.

Ben-Ari, N. (1988). The ambivalent case of repetitions in literary translation. Avoiding repetitions: A "universal" of translation. *Meta*, 43(1), 68–78.

Beňuš, S., Gravano, A., and Hirschenberg, J. (2011). Pragmatic aspects of temporal accommodation in turn-taking. *Journal of Pragmatics*, 43, 3001–3027.

Benzitoun, C., Debaisieux, J.-M., and Deulofeu, H.-J. (2016). Le projet ORFEO : un corpus d'étude pour le français contemporain. *Corpus*, 15, 1–18.

Bernardini, S. (2004). Corpora in the classroom: An overview and some reflections on future developments. In *How to Use Corpora in the Language Classroom*, Sinclair, J. (ed.). John Benjamins, Amsterdam, 15–36.

Bernicot, J., Volckaert-Legrier, O., Goumi, A., and Bert-Erboul, A. (2012). Forms and functions of SMS messages: A study of variations in a corpus written by adolescents. *Journal of Pragmatics*, 44, 1701–1715.

Berthelet, R. (2015). Googling Toubon. Testing the effects of institutional French language purism. In *Change of Paradigms – New Paradoxes: Recontextualizing Language and Linguistics*, Daems, J., Zenner, E., Heylen, K., Speelman, D., and Cuyckens, H. (eds). de Gruyter Mouton, Berlin.

Bertrand, R., Blache, P., Espesser, R., Ferré, G., Meunier, C., Priego-Valverde, B., and Rauzy, S. (2008). Le CID – Corpus of Interactional Data – annotation et exploitation multimodale de parole conversationnelle. *Traitement automatique des langues*, 49(3), 1–30.

Biber, D. (1993). Representativeness in corpus design. *Literary and Linguistic Computing*, 8(4), 243–257.

Biber, D. and Conrad, S. (2019). *Register, Genre and Style*, 2nd edition. Cambridge University Press, Cambridge.

Biber, D. and Reppen, R. (2002). What does frequency have to do with language teaching? *Studies in Second Language Acquisition*, 24, 199–208.

Blum-Kulka, S. (1986). Shifts of cohesion and coherence in translation. In *Interlingual and Intercultural Communication*, House, J. and Blum-Kulka, S. (eds). Narr, Tübigen, 17–35.

Boré, C. and Elalouf, M.-L. (2017). Deux étapes dans la construction de corpus scolaires : problèmes récurrents et perspectives nouvelles. *Corpus*, 16, 31–63.

Branca-Rosoff, S., Fleury, S., Lefeuvre, F., and Pires, M. (2012). Discours sur la ville. Présentation du Corpus de Français Parlé Parisien des années 2000 (CFPP2000) [Online]. Available: http://www.univ-paris3.fr/discours-sur-la-ville-corpus-de-francais-parle-parisien-des-annees-2000-cfpp2000--39737.kjsp?RH=1505727285324.

Braun, S. (2005). From pedagogically relevant corpora to authentic language learning contents. *ReCALL*, 17, 47–64.

Brezina, V. (2018). *Statistics in Corpus Linguistics. A Practical Guide*. Cambridge University Press, Cambridge.

Brown, R. (1973). *A First Language: The Early Stages*. Harvard University Press, Cambridge.

Bull, C., Asfiandy, D., Gledson, A., Mellor, J., Couth, S., Stringer, G., Rayson, P., Sutcliffe, A., Keane, J., Zeng, X., Burns, A., Leroi, I., Ballard, C., and Sawyer, P. (2016). Combining data mining and text mining for detection of early stage dementia: The SAMS framework. In *Proceedings of RaPID–2016*, Kokkadinis, D. (ed.). Portoroz, Slovenia, 35–40.

Bunt, H., Petukhova, V., Traum, D., and Alexandersson, J. (2012). Dialogue act annotation with the ISO 24617–2 standard. In *Multimodal Interaction with W3C Standards: Towards Natural User Interfaces to Everything*, Dahl, D. (ed.). Springer, Dordrecht, 109–135.

Buysse, L. (2020). "It was a bit stressy as well actually". The pragmatic markers *actually* and *in fact* in spoken learner English. *Journal of Pragmatics*, 156, 28–40.

Bybee, J. (2001). *Phonology and Language Use*. Cambridge University Press, Cambridge.

Cartoni, B., Zufferey, S., and Meyer, T. (2013a). Annotating the meaning of discourse connectives by looking at their translation: The translation spotting technique. *Dialogue and Discourse*, 4(2), 65–86.

Cartoni, B., Zufferey, S., and Meyer, T. (2013b). Using the Europarl corpus for linguistic research. *Belgian Journal of Linguistics*, 27, 23–42.

Čermáková, A. (2015). Repetition in John Irving's novel A Widow for One Year. *International Journal of Corpus Linguistics*, 20(3), 355–377.

Cheng, W. (2010). What can a corpus tell us about language teaching? In *The Routledge Handbook of Corpus Linguistics*, O'Keeffe, A. and McCarthy, M. (eds). Routledge, Abingdon-on-Thames, 319–332.

Chesterman, A. (1998). *Contrastive Functional Analysis*. John Benjamins, Amsterdam.

Chevrot, J.-P., Dugua, C., Harnois-Delpiano, M., Siccardi, A., and Spinelli, E. (2013). Liaison acquisition: Debates, critical issues, future research. *Language Science*, 39, 83–94.

Cohen, J. (1960). A coefficient of agreement for nominal scales. *Educational and Psychological Measurement*, 20, 37–46.

Cohen, J. (1988). *Statistical Power Analysis for the Behavioral Sciences*, 2nd edition. Lawrence Erlbaum Associates, Hillsdale.

Cole, J., Hasegawa-Johnson, M., Loehr, D., Van Guilder, L., Reetz, H., and Frisch, S. (2011). Corpora, databases, and internet resources: Corpus phonology with speech resources using the internet for collecting phonological data speech manipulation, synthesis, and automatic recognition in laboratory phonology phonotactic patterns in lexical corpora. In *The Oxford Handbook of Laboratory Phonology*, Cohn, A., Fougeron, C., and Huffman, M. (eds). Oxford University Press, Oxford, 1–47.

Cotterhill, J. (2010). How to use corpus linguistics in forensic linguistics? In *The Routledge Handbook of Corpus Linguistics*, O'Keeffe, A. and McCarthy, M. (eds). Routledge, Abingdon-on-Thames, 578–590.

Cougnon, L.-A. (2015). *Langage et sms: une étude internationale des pratiques actuelles*. Presses Universitaires de Louvain, Louvain-la-Neuve.

Cougnon, L.-A., Maskens, L., Roekhaut, S., and Fairon, C. (2017). Social media, spontaneous writing and dictation. Spelling variation. *Journal of French Language Studies*, 27, 309–327.

Coulthard, M., Johnson, A., and Wright, D. (2017). *An Introduction to Forensic Linguistics: Language in Evidence*, 2nd edition. Routledge, Abingdon-on-Thames.

Cresti, E. and Moneglia, M. (2005). *C-ORAL-ROM: Integrated Reference Corpora for Spoken Romance Languages*. John Benjamins, Amsterdam.

Crible, L. (2018). *Discourse Markers and (Dis)fluency: Forms and Functions Across Languages and Registers*. John Benjamins, Amsterdam.

Cummins, S. and Desjardins, I. (2002). A case study in lexical research for translation. *International Journal of Lexicography*, 15(2), 139–156.

Danlos, L., Antolinos-Basso, D., Braud, C. and Roze, C. (2012). Vers le FDTB : French discourse tree bank. *Proceedings of TALN 2012, 19éme conférence sur le Traitement automatique des langues naturelles*. Grenoble, 471–478.

Da Silva Genest, C. and Masson, C. (2019). Corpus et pathologies du language: du recueil à l'analyse de données pour une linguistique clinique et appliquée. *Corpus*, 19, 1–7.

De Cat, C. and Plunkett, B. (2002) "QU' est ce qu' i(l) dit, celui+Là ?" : notes méthodologiques sur la transcription d'un corpus francophone. In *Romanistische Korpuslinguistik: Korpora und gesprochene Sprache/Romance Corpus Linguistics: Corpora and Spoken Language*, Pusch, C. and Raible, W. (eds). Narr, Tübingen.

DeForest, M. and Johnson, E. (2001). The density of Latinate words in the speeches of Jane Austen's characters. *Literary and Linguistic Computing*, 16(4), 389–401.

Degand, L. (2004). Contrastive analyses, translation and speaker involvement: The case of *puisque* and *aangezien*. In *Language, Culture and Mind*, Achard, M. and Kemmer, S. (eds). The University of Chicago Press, Chicago, 251–270.

Degand, L. and Fagard, B. (2011). Alors between discourse and grammar: The role of syntactic position. *Functions of Language*, 18(1), 29–56.

Degand, L. and Fagard, B. (2012). Competing connectives in the causal domain: French *car* and *parce que*. *Journal of Pragmatics*, 44, 154–168.

Degen, J. (2015). Investigating the distribution of *some* (but not *all*) implicatures using corpora and web-based methods. *Semantics and Pragmatics*, 8, 1–55.

Detey, S., Lyche, C., Tchobanov, A., Durand, J., and Laks, B. (2009). Ressources phonologiques au service de la didactique de l'oral : le projet PFC-EF. *Mélanges CRAPEL*, 31, 223–236.

Deutscher, G. (2011). *Through the Language Glass: Why the World Looks Different in Other Languages*. Arrow Books, London.

Diessel, H. (2009). Corpus linguistics in first language acquisition. In *Corpus Linguistics: An International Handbook*, Lüdeling, A. and Kytö, M. (eds). de Gruyter Mouton, Berlin, 2, 1197–1212.

Dister, A., Francard, M., Hambye, P., and Simon, A.-C. (2009). Du corpus à la banque de données. Du son, des textes et des métadonnées. L'évolution de banque de données textuelles orales VALIBEL (1989–2009). *Cahiers de linguistique*, 33(2), 113–129.

Dostie, G. (2012). Le Corpus de français parlé au Québec (CFPQ) et la langue des conversations familières : exemple de mise à profit des données à partir d'un examen lexico-sémantique de la séquence *je ne sais pas*. *Corpus*, 15, 1–13.

Dupont, M. and Zufferey, S. (2017). Methodological issues in the use of parallel directional corpora: A case study with English and French concessive connectives. *International Journal of Corpus Linguistics*, 22(2), 270–297.

Dürscheid, C. and Stark, E. (2011). SMS4science: An international corpus-based texting project and the specific challenges for multilingual Switzerland. In *Digital Discourse: Language in the New Media*, Thurlow, C. and Mroczek, K. (eds). Oxford University Press, Oxford, 299–320.

Evers-Vermeul, J. and Sanders, T. (2011). Discovering domains: On the acquisition of causal connectives. *Journal of Pragmatics*, 43(6), 1645–1662.

Evison, J. (2010). What are the basics of analyzing a corpus? In *The Routledge Handbook of Corpus Linguistics*, O'Keeffe, A. and McCarthy, M. (eds). Routledge, Abingdon-on-Thames, 122–135.

Fabre, C. and Bourigault, D. (2008). Exploiter des corpus annotés syntaxiquement pour observer le continuum entre arguments et circonstants. *Journal of French Language Studies*, 18, 87–102.

Fairon, C., Klein, S., and Paumier, S. (2006). *Le langage SMS: étude d'un corpus informatisé à partir de l'enquête "faites don de vos SMS à la science"*. Presses Universitaires de Louvain, Louvain-la-Neuve.

Ferguson, A., Craig, H., and Spencer, E. (2009). Exploring the potential for corpus-based research in speech-language pathology. In *Selected Proceedings of the 2008 HCSNet Workshop on Designing the Australian National Corpus*, Haugh, M., Burridge, K., Mulder, J., and Peters, P. (eds). Cascadilla Proceedings Project, Somerville, 30–36.

Field, A. (2017). *Discovering Statistics Using IBM SPSS Statistics*, 5th edition. Sage Publications, Thousand Oaks.

Forsberg Lundell, F., Bartning, I., Engel, H., Gudmundon, A., Hancock, V., and Lindqvist, C. (2014). Beyond advanced stages in high-level spoken L2 French. *Journal of French Language Studies*, 24, 255–280.

Fradin, B. (1997). Esquisse d'une sémantique de la préfixation en anti-. *Recherches linguistiques de Vincennes*, 26, 87–112.

Fraser, K. and Hirst, G. (2016). Detecting semantic changes in Alzheimer's disease with vector space models. In *Proceedings of RaPID–2016*, Kokkadinis, D. (ed.). Portoroz, Slovenia, 1–8.

Gardent, C. and Manuélian, H. (2005). Création d'un corpus annoté pour le traitement des descriptions définies. *L'objet*, 8, 1–15.

Gardner-Chloros, P. and Secova, M. (2018). Grammatical change in Paris French: *In situ* question words in embedded contexts. *Journal of French Language Studies*, 28, 181–207.

Garside, R., Hutchinson, J., Leech, G., McEnery, A., and Oakes, M. (1994). The exploittation of parallel corpora in Projects ET10/63 and CRATER. *New Methods in Language Processing*. UMIST, Manchester, 108–115.

Gast, V. (2012). Contrastive analysis: Theories and methods. In *Dictionaries of Linguistics and Communication Science: Linguistic Theory and Methodology*, Kortmann, B. and Kabatek, J. (eds). de Gruyter Mouton, Berlin.

Gillioz, C. and Zufferey, S. (forthcoming). *Introduction to Experimental Linguistics*. ISTE Ltd, London, and Wiley, New York.

Gilquin, G. (2008). Causative "make" and "faire": A case of mismatch. In *Current Trends in Contrastive Linguistics: Functional and Cognitive Perspectives*, de los Ángeles Gómez González, M., Lachlan Mackenzie, J., and González Álvarez, E.M. (eds). John Benjamins, Amsterdam, 177–201.

Gilquin, G. (2015). From design to collection of learner corpora. In *The Cambridge Handbook of Learner Corpus Research*, Granger, G., Gilquin, G., and Meunier, F. (eds). Cambridge University Press, Cambridge, 9–34.

Gilquin, G. and Gries, S. (2009). Corpora and experimental methods: A state-of-the-art review. *Corpus Linguistics and Linguistic Theory*, 5, 1–26.

Granger, S. (2008). Learner corpora in foreign language education. In *Encyclopedia of Language and Education 24*, Van Deusen-Scholl, N. and Hornberger, N. (eds). Springer, New York, 337–351.

Granger, S., Gilquin, G., and Meunier, F. (2015). *The Cambridge Handbook of Learner Corpus Research*. Cambridge University Press, Cambridge.

Grice, H.P. (1989). *Studies in the Way of Words*. Harvard University Press, Cambridge.

Gries, S. (2008). Dispersions and adjusted frequencies in corpora. *International Journal of Corpus Linguistics*, 13(4), 403–437.

Gries, S. (2013). *Statistics for Linguistics with R: A Practical Introduction*. Walter de Gruyter, Berlin.

Groupe ICOR (Bert, M., Bruxelles, S., Étienne, C., Mondada, L., and Traverso, V.) (2008). Tool-assisted analysis of interactional corpora: "*viola*" in the CLAPI database. *Journal of French Language Studies*, 18, 121–145.

Guillot-Barbance, C., Heiden, S., and Lavrentiev, A. (2017). Base de français médiéval : une base de référence de sources médiévales ouverte et libre au service de la communauté scientifique. *Diachroniques*, 7, 168–184.

Habert, B., Nazarenko, A., and Salem, A. (1997). *Les linguistiques de corpus*. Armand Colin, Paris.

Hanks, P. (2012). The corpus revolution in lexicography. *International Journal of Lexicography*, 25(4), 398–436.

Hathout, N., Plénat, M., and Tanguy, L. (2003). Enquête sur les dérivés en *-able*. *Cahiers de grammaire*, 28, 49–90.

Hathout, N., Montermini, F., and Tanguy, L. (2008). Extensive data for morphology: Using the World Wide Web. *Journal of French Language Studies*, 18, 67–85.

Hausmann, F. and Gorbahn, A. (1989). Review article: COBUILD and LDOCE II – A comparative review. *International Journal of Lexicography*, 2, 44–56.

Hoff-Ginsberg, E. (1991). Mother–child conversation in different social classes and communicative social settings. *Child Development*, 62, 782–796.

Howell, D. (1998). *Méthodes statistiques en sciences humaines*. De Boeck, Brussels.

Howes, C., Lavelle, M., Healey, P., Hough, J., and McCabe, R. (2017). Disfluencies in dialogues with patients with schizophrenia. *Proceedings of the 39th Annual Meeting of the Cognitive Science Society*. London.

Hunston, S. (2002). *Corpora in Applied Linguistics*. Cambridge University Press, Cambridge.

Ide, N. and Pustjevosky, J. (2017). *Handbook of Linguistic Annotations*, 2 volumes. Springer, Dordrecht.

Ilisei, I., Inkpen, D., Corpas Pastor, G., and Mitkov, R. (2010). Identification of translationese: A machine learning approach. In *Computational Linguistics and Intelligent Text Processing*, Gelbukh, A. (ed.). Springer, Berlin, 503–511.

Jaszczolt, K. (2003). On translating "what is said": Tertium comparationis in contrastive semantics and pragmatics. In *Meaning through Language Contrast*, Jaszczolt, K. and Turner, K. (eds). John Benjamins, Amsterdam, 441–462.

Johansson, S. (1998). On the role of corpora in cross-linguistic research. In *Corpora and Cross-linguistic Research: Theory, Method and Case Studies*, Johansson, S. and Oksefjell, S. (eds). Rodopi, Amsterdam, 3–24.

Johansson, S. (2007). *Seeing through Multilingual Corpora: On the Use of Corpora in Contrastive Studies*. John Benjamins, Amsterdam.

Jones, S. (2002). *Antonymy: A Corpus-based Perspective*. Routledge, Abingdon-on-Thames.

Jucker, A. (2009). Speech act research between armchair, field and laboratory: The case of compliments. *Journal of Pragmatics*, 41, 1611–1635.

Karrass, J., Braungart-Rieker, J., Mullins, J. and Lefever, J. (2002). Processes in language acquisition: The roles of gender, attention, and maternal encouragement of attention over time. *Journal of Child Language*, 29, 519–543.

Kennedy, G. (2003). Amplifier collocations in the British National Corpus: Implications for English language teaching. *TESOL Quarterly*, 37(3), 467–487.

Kenning, M.-M. (2010). What are parallel and comparable corpora and how we use them? In *The Routledge Handbook of Corpus Linguistics*, O'Keeffe, A. and McCarthy, M. (eds). Routledge, Abingdon-on-Thames, 487–500.

Kidd, E., Donnelly, S., and Christiansen, M. (2017). Individual differences in language acquisition and processing. *Trends in Cognitive Sciences*, 22(2), 154–169.

Kilgarriff, A., Baisa, V., Bušta, J., Jakubíček, M., Kovář, V., Michelfeit, J., Rychlý, P., and Suchomel, V. (2014). The Sketch Engine: Ten years on. *Lexicography*, 1, 7–36.

Koehl, A. and Lignon, S. (2014). Property nouns with *-ité* and *-itude*: Formal alternation and morphopragmatics of the sad-itude of the A*ité*N. *Morphology*, 24, 351–376.

Koehn, P. (2005). Europarl: A parallel corpus for statistical machine translation. *Proceedings of the 10th Machine Translation Summit*. Phuket, 79–86.

Koester, A. (2010). Building a small specialised corpora. In *The Routledge Handbook of Corpus Linguistics*, O'Keeffe, A. and McCarthy, M. (eds). Routledge, Abingdon-on-Thames, 66–79.

Koppel, M., Argamon, S., and Shimoni, A. (2002). Automatically categorizing written texts by author gender. *Literary and Linguistic Computing*, 17(4), 401–412.

Kredens, K. and Coulthard, M. (2012). Corpus linguistics in authorship identification. In *The Oxford Handbook of Language and the Law*, Solan, L. and Tiersma, P. (eds). Oxford University Press, Oxford, 504–516.

Krezeszowski, T. (1990). *Contrasting Languages: The Scope of Contrastive Linguistics*. Walter de Gruyter, Berlin.

Krippendorff, K. (1980). *Content Analysis: An Introduction to its Methodology*. Sage Publications, Beverly Hills.

Kuebler, S. and Zinsmeister, H. (2014). *Corpus Linguistics and Linguistically Annotated Corpora*. Bloomsbury, London.

Kyratzis, A., Guo, J., and Ervin-Tripp, S. (1990). Pragmatic conventions influencing children's use of causal constructions in natural discourse. *Proceedings of the 16th Annual Meeting of the Berkeley Linguistics Society*. Berkeley, Knightsbridge, 205–214.

Lacheret-Dujour, A., Kahane, S., and Pietrandrea, P. (2019). *A Prosodic and Syntactic Treebank for Spoken French*. John Benjamins, Amsterdam.

Ladd, D. (2008). *Intonational Phonology*, 2nd edition. Cambridge University Press, Cambridge.

Lado, R. (1957). *Linguistics Across Cultures: Applied Linguistics for Language Teachers*. University of Michigan Press, Ann Arbor.

Lauridsen, K. (1996). Text corpora and contrastive linguistics: Which type of corpus for which type of analysis? In *Languages in Contrast: Papers from a Symposium on Text-based Cross-linguistic Studies*, Aijmer, K., Altenberg, B., and Johansson, M. (eds). Lund University Press, Lund, 63–71.

Laviosa-Braithwaite, S. (1997). Investigating simplification in an English comparable corpus of newspaper articles. In *Transferre necesse est. Proceedings of the 2nd International Conference on Current Trends in Studies of Translation and Interpreting*, Klaudy, K. and Kohn, J. (eds). Scholastica, Budapest, 531–540.

Laviosa-Braithwaite, S. (2009). Universals of translation. In *Routledge Encyclopedia of Translation Studies*, Baker, M. and Saldanha, G. (eds). Routledge, Abingdon-on-Thames, 288–291.

Lee, D. (2010). What corpora are available? In *The Routledge Handbook of Corpus Linguistics*, O'Keeffe, A. and McCarthy, M. (eds). Routledge, London, 107–121.

Lefer, M.-A. and Grabar, N. (2015). Super-creative and over-bureaucratic: A cross-genre corpus-based study on the use and translation of evaluative prefixation in TED talks and EU parliamentary debates. *Across Languages and Cultures*, 16, 187–208.

Levi, J. (1993). Evaluating jury comprehension of the Illinois capital sentencing instruction. *American Speech*, 68, 20–49.

Levinson, S. (2000). *Presumptive Meanings: The Theory of Generalized Conversational Implicature*. MIT Press, Cambridge.

Lin, Y., Michel, J.-B., Lieberman Aiden, E., Orwant, J., Brockman, W., and Petrov, S. (2012). Syntactic annotation for the Google books Ngram corpus. *Proceedings of the 50th Meeting of the Association for Computational Linguistics*. Jeju, 169–174.

Litosseliti, L. (2018). *Research Methods in Linguistics*, 2nd edition. Bloomsbury, London.

Lüdeling, A. and Hirschmann, H. (2015). Error annotation systems. In *The Cambridge Handbook of Learner Corpus Research*, Granger, S., Gilquin, G., and Meunier, F. (eds). Cambridge University Press, Cambridge, 135–159.

Lyster, R. (2006). Predictability of French gender attribution: A corpus analysis. *Journal of French Language Studies*, 16, 69–92.

MacWhinney, B. (2000). *The CHILDES Project: Tools for Analyzing Talk. Vol. II: The Database*, 3rd edition. Lawrence Erlbaum, Hillsdale.

MacWhinney, B. and Snow, C. (1985). The child language data exchange system. *Journal of Child Language*, 12(2), 271–296.

Mahleberg, M. (2014). Corpus stylistics. In *The Routledge Handbook of Stylistics*, Burke, M. (ed.). Routledge, Abingdon-on-Thames, 378–392.

Mahleberg, M. and McIntyre, D. (2011). A case for corpus stylistics: Ian Fleming's *Casino Royale*. *English Text Construction*, 4(2), 204–227.

Mauranen, A. (1999). Will "translationese" ruin a contrastive study? *Languages in Contrast*, 2(2), 161–185.

Mauranen, A. and Kujamäki, P. (2004). *Translation Universals: Do They Exist?* John Benjamins, Amsterdam.

McCarten, J. (2010). Corpus-informed course book design. In *The Routledge Handbook of Corpus Linguistics*, O'Keeffe, A. and McCarthy M. (eds). Routledge, Abingdon-on-Thames, 413–427.

McEnery, T. and Hardie, A. (2012). *Corpus Linguistics*. Cambridge University Press, Cambridge.

McEnery, T. and Wilson, A. (2001). *Corpus Linguistics*. Edinburgh University Press, Edinburgh.

McEnery, T., Xiao, R., and Tono, Y. (2005). *Corpus-based Language Studies: An Advanced Resource Book*. Routledge, Abingdon-on-Thames.

McEwan, C., Ounis, I., and Ruthven, I. (2002). Building bilingual dictionaries from parallel web documents. *Proceedings from Advances in Information Retrieval, 24th BCS-IRSG European Colloquium on IR Research*. Glasgow, 303–323.

McIntyre, D. and Walker, B. (2010). How can corpora be used to explore the language of poetry and drama? In *The Routledge Handbook of Corpus Linguistics*, O'Keeffe, A. and McCarthy, M. (eds). Routledge, Abingdon-on-Thames, 516–530.

McWhorter, J. (2016). *The Language Hoax: Why the World Looks the Same in Every Language*. Oxford University Press, Oxford.

Meinschaeffer, J., Bonifer, S., and Frisch, C. (2015). Variable and invariable liaison in a corpus of spoken French. *Journal of French Language Studies*, 25, 367–396.

Michel, J.-B., Shen, Y., Aiden, A., Veres, A., Gray, M., The Google Books Team, Pickett, J., Hoiberg, D., Clancy, D., Norvig, P., Orwant, J., Pinker, S., Nowak, M., and Lieberman, E. (2011). Quantitative analysis of culture using millions of digitized books. *Science*, 331(6014), 176–182.

Mikhailov, M. and Cooper, R. (2016). *Corpus Linguistics for Translation and Contrastive Studies: A Guide for Research*. Routledge, Abingdon-on-Thames.

Murray, J. (1997). Connectives and narrative text: The role of continuity. *Memory and Cognition*, 25, 227–236.

Nelson, M. (2010). Building a written corpus: What are the basics? In *The Routledge Handbook of Corpus Linguistics*, O'Keeffe, A. and McCarthy, M. (eds). Routledge, Abingdon, 53–65.

Noël, D. (2003). Translations as evidence for semantics: An illustration. *Linguistics*, 41(4), 757–785.

Oakes, M. (2009). Corpus linguistics and stylometry. In *Corpus Linguistics: An International Handbook. Vol. 2*, Lüdeling, A. and Kytö, M. (eds). de Gruyter Mouton, Berlin.

Ogiermann, E. (2009). Politeness and in-directness across cultures: A comparison of English, German, Polish and Swedish requests. *Journal of Politeness Research*, 5, 189–216.

O'Keeffe, A. and McCarthy, M. (2010). *The Routledge Handbook of Corpus Linguistics*. Routledge, Abingdon-on-Thames.

Ortega, L. (2014). *Understanding Second Language Acquisition*, 2nd edition. Routledge, Abingdon.

Ortolang (2019). Débats parlementaires sur l'Europe à la House of Commons (1998–2015). Ortolang (Open Resources and TOols for LANGuage) [Online]. Available: https://www.ortolang.fr/market/corpora/uk-parl.

Ostler, N. (2009). Existing corpora. In *Corpus Linguistics: An International Handbook. Vol. 2*, Lüdeling, A. and Kytö, M. (eds). de Gruyter Mouton, Berlin, 383–483.

Ozdowska, S. (2009). Données bilingues pour la TAS français-anglais : impact de la langue source et direction de traduction originales sur la qualité de la traduction. *Proceedings of traitement automatique des langues naturelles*. Senlis.

Panckhurst, R., Détrie, C., Lopez, C., Moïse, C., Roche, M., and Verine, B. (2013). Sud4science, de l'acquisition d'un grand corpus de SMS en français à l'analyse de l'écriture de SMS. *Epistémè*, 7, 107–138.

Parisse, C. and Le Normand, M.-T. (2000). How children build their morphosyntax: The case of French. *Journal of Child Language*, 27, 267–292.

Parisse, C. and Morgenstern, A. (2012). The unfolding of the verbal system in French children's speech between 18 and 36 months. *Journal of French Language Studies*, 22, 95–114.

Patar, A., Grabar, N., and de Mulder, W. (2015). Étude diachronique du conditionnel passé ou l'origine de la contrefactualité. *Journal of French Language Studies*, 25, 189–211.

Pierrel, J.-M., Dendien, J., and Bernard, P. (2004). Le TLFi ou Trésor de la Langue Française informatisé. *Actes de EURALEX 2004*. Lorient, France, 165–170.

Pit, M. (2007). Cross-linguistic analyses of backward causal connectives in Dutch, German and French. *Languages in Contrast*, 7(1), 53–82.

Planchon, C. (2018). Anglicismes dans la presse écrite: le bilinguisme de milieu peut-il expliquer l'anglicisation ? *Journal of French Language Studies*, 28, 43–66.

Prasad, R., Dinesh, N., Lee, A., Miltsakaki, E., Robaldo, L. Joshi, A., and Webber, B. (2008). The Penn Discourse TreeBank 2.0. *Proceedings of the 6th International Conference on Language Resources and Evaluation*. Marrakech, 2961–2968.

Racine, I. (2014). Une approche par corpus de la liaison chez les apprenants hispanophones de français langue étrangère : quelles conséquences pour l'enseignement du FLE ? *Flambeau*, 40, 18–37.

Racine, I. and Detey, S. (2017). Pour un renouvellement de l'enseignement de la liaison en FLE au regard des corpus : défis d'apprentissage et usages contemporains. *Journal of French Language Studies*, 27, 87–99.

Reppen, R. (2010a). *Using Corpora in the Language Classroom*. Cambridge University Press, Cambridge.

Reppen, R. (2010b). Building a corpus: What are the key considerations? In *The Routledge Handbook of Corpus Linguistics*, O'Keeffe, A. and McCarthy, M. (eds). Routledge, Abingdon-on-Thames, 31–37.

Rice, M. (2010). Mean length of utterance levels in 6-months intervals for children 3 to 9 years with and without language impairment. *Journal of Speech and Language Hearing Research*, 53(2), 333–349.

Rowland, C. (2013). *Understanding Language Acquisition*. Routledge, Abingdon-on-Thames.

Roy, B., Frank, M., and Roy, D. (2012). Relating activity contexts to early word learning in dense longitudinal data. *Proceedings of the 34th Annual Meeting of the Cognitive Science Society*. Sapporo, Japan.

Royle, P. and Stine, I. (2013). The French noun phrase in preschool children with SLI: Morphosyntactic and error analysis. *Journal of Child Language*, 40, 945–970.

Roze, C., Danlos, L., and Muller, P. (2012). LEXCONN: A French lexicon of discourse connectives. *Discours*, 10, 1–15.

Rühlemann, C. (2018). *Corpus Linguistics for Pragmatics. A Guide for Research*. Routledge, Abingdon-on-Thames.

Sampson, G. (2001). *Empirical Linguistics*. Continuum, London.

Santini, M. (2004). A shallow approach to syntactic feature extraction for genre classification. *Proceedings from the 7th Annual Research Colloquium for the UK Special Interest Group for Computational Linguistics (CLUK)*. University of Birmingham, Birmingham, 207–214.

Schwab, S. and Avanzi, M. (2015). Regional variation and articulation rate in French. *Journal of Phonetics*, 48, 96–105.

Seiger-Gardner, L. and Almodovar, D. (2017). Language production approaches to child language disorders. In *Handbook of Child Language Disorders*, Schwarz, R. (ed.). Routledge, Abingdon-on-Thames, 465–487.

Semino, E. and Short, M. (2004). *Corpus Stylistics: Speech, Writing and Thought Presentation in a Corpus of English Writing*. Routledge, Abingdon-on-Thames.

Sharoff, S., Rapp, R., Zweigenbaum, P., and Fung, P. (2016). *Building and Using Comparable Corpora*. Springer, Berlin.

Siepmann, D. (2015). Dictionaries and spoken language: A corpus-based review of French dictionaries. *International Journal of Lexicography*, 28(2), 139–168.

Siepmann, D., Bürgel, C., and Diwersy, S. (2016). Le corpus de référence du français contemporain (CRFC) : un corpus massif du français largement diversifié par genres. *SHS Web of Conferences*, 27(11002), 1–13.

Simon, A.-C. and Degand, L. (2007). Connecteurs de causalité, implication du locuteur et profils prosodiques : le cas de *car* et de *parce que*. *Journal of French Language Studies*, 17(3), 323–341.

Sinclair, J. (1991). *Corpus, Concordance, Collocation*. Oxford University Press, Oxford.

Sinclair, J. (1992). The automatic analysis of text corpora. In *Directions in Corpus Linguistics: Proceedings of the Nobel Symposium 82, Stockholm*, Startvik, J. (ed.). de Gruyter Mouton, The Hague, 379–397.

Sinclair, J. (2004). *How to Use Corpora in Language Teaching*. John Benjamins, Amsterdam.

Spooren, W. and Degand, L. (2010). Coding coherence relations: Reliability and validity. *Corpus Linguistics and Linguistic Theory*, 6(2), 241–266.

Steffens, M. (2018). Antonymic discourse functions and manipulation: A corpus analysis of present-day French. *Corpus Pragmatics*, 2, 313–332.

Steinberger, R., Pouliquen, B., Widiger, A., Ignat, C., Erjavec, T., Tufis, D., and Varga, D. (2006). The JRC-Acquis: A multilingual aligned parallel corpus with 20+ languages. *Proceedings of the 5th International Conference on Language Resources and Evaluation (LREC'2006)*. Genoa.

Stern, C. and Stern, W. (1907). *Die Kindersprache: Eine psychologische und sprachtheoretische Untersuchung*. Barth, Leipzig.

Stosic, D., Marjanović, S., and Miletic, A. (2018). ParCoGLiJe : corpus parallèle pour l'étude des grands classiques de la littérature de jeunesse. *Poster présenté à la Journée d'étude CORLI Traitements et standardisation des corpus multimodaux et web 2.0*. Paris.

Stubbs, M. (1996). *Text and Corpus Analysis*. Blackwell, Oxford.

Szudarski, P. (2017). *Corpus Linguistics for Vocabulary: A Guide for Research*. Routledge, Abingdon.

Tager-Flusberg, H., Calkins, S., Nolin, T., Baumberger, T., Anderson, M., and Chadwick-Dias, A. (1990). A longitudinal study of language acquisition in autistic and Down syndrome children. *Journal of Autism and Developmental Disorders*, 20, 1–21.

Taylor, A., Marcus, M., and Santorini, B. (2003). The Penn Treebank: An overview. In *Building and Using Parsed Corpora*, Abeillé, A. (ed.). Springer, Dordrecht, 5–22.

Thuilier, J., Fox, G., and Crabé, B. (2010). Approche quantitative en syntaxe : l'exemple de l'alternance de position de l'adjectif épithète en français. *Actes de traitement automatique des langues naturelles*, September, 1–11.

Tiersema, P. (2002). The language and law of products warning. In *Language in the Legal Process*, Cotterill, J. (ed.). Palgrave, London, 54–71.

Timmis, I. (2015). *Corpus Linguistics for ELT: Research and Practice*. Routledge, Abingdon.

Tognini-Bonelli, E. (2010). The evolution of corpus linguistics. In *The Routledge Handbook of Corpus Linguistics*, O'Keeffe, A. and McCarthy, M. (eds). Routledge, Abingdon, 14–28.

Tomasello, M. (2003). *Constructing a Language: A Usage-based Theory of Language Acquisition*. Harvard University Press, Cambridge.

Tomasello, M. and Stahl, D. (2004). Sampling children's spontaneous speech: How much is enough? *Journal of Child Language*, 31, 101–121.

Traverso, V. (2019). Demander de l'aide à la permanence d'accès aux droits d'un centre social : modalité de construction des requêtes. *Journal of French Language Studies*, 29, 113–136.

Truan, N. (2016). Parliamentary debates on Europe at the House of Commons (1998–2015) Ortolang (Open Resources and TOols for LANGuage) [Online]. Available: https://hdl.handle.net/11403/uk–parl.

Véronis, J. and Langlais, P. (2000). Evaluation of parallel text alignment systems: The arcade project. In *Parallel Text Processing*, Véronis, J. (ed.). Kluwer, Alphen-sur-le-Rhin, 369–388.

Verwimp, L. and Lahousse, K. (2017). Definite *il y a*-clefts in spoken French. *Journal of French Language Studies*, 27, 263–290.

Walter, E. (2010). Using corpora to write dictionaries. In *The Routledge Handbook of Corpus Linguistics*, O'Keeffe, A. and McCarthy, M. (eds). Routledge, Abingdon-on-Thames, 428–443.

Wild, K., Kilgarriff, A., and Tugwell, D. (2012). The Oxford's children corpus: Using a children's corpus in lexicography. *International Journal of Lexicography*, 26(2), 190–218.

Wright, H.H. (2011). Discourse in aphasia: An introduction to current research and future directions. *Aphasiology*, 25, 1283–1285.

Wynne, M. (2005). Archiving, distribution and preservation. In *Developing Linguistic Corpora: A Guide to Good Practice*, Wynne, M. (ed.). Oxbow Books, Oxford, 71–78.

Zipf, G. (1932). *Selective Studies of the Principle of Relative Frequency in Language*. Harvard University Press, Cambridge.

Zufferey, S. (2010). *Lexical Pragmatics and Theory of Mind: The Acquisition of Connectives*. John Benjamins, Amsterdam.

Zufferey, S. and Cartoni, B. (2012). English and French causal connectives in contrast. *Languages in Contrast*, 12(2), 232–250.

Zufferey, S. and Cartoni, B. (2014). A multifactorial analysis of explicitation in translation. *Target*, 26(3), 361–384.

Zufferey, S., Moeschler, J., and Reboul, A. (2019). *Implicatures*. Cambridge University Press, Cambridge.

Webography

Chapter 5

Corpus gathering (accessed 30.06.2019)

Collection de corpus oraux numériques (COCOON): https://cocoon.huma-num.fr/exist/crdo/.

Corpus d'étude du français contemporain: https://www.projet-orfeo.fr/.

Learner corpora around the World: https://uclouvain.be/en/research-institutes/ilc/cecl/learner-corpora-around-the-world.html.

Opus: http://opus.nlpl.eu/.

Sketch Engine: https://www.sketchengine.eu/.

Corpus dissemination sites (accessed 30.06.2019)

European Language Resources Association (ELRA): http://www.elra.info/en/.

Linguistic Data Consortium (LDC): https://www.ldc.upenn.edu/.

Plateforme Ortolang: https://www.ortolang.fr/.

Sites for downloading and learning how to use corpus consultation tools (accessed 30.06.2019)

AntConc: http://www.laurenceanthony.net/software/antconc/.

CLAN: http://dali.talkbank.org/clan/.

ParaConc: http://paraconc.com/.

The Programming Historian: https://programminghistorian.org/en/lessons/corpus-analysis-with-antconc.

Summary table of the corpora presented (consulted 30.06.2019)

Corpus name	Type of corpus	Number of words	Availability	Access
Base de Français Médiéval	Diachronic writing	4,000,000 (BFM)	Online	http://bfm.ens-lyon.fr/
Corpus Français de l'Université de Leipzig	Varied writing	425,000,000	Online	http://wortschatz.uni-leipzig.de/de

Frantext	Literary writing	250,000,000	Subscription or partially on Ortolang	http://www.cnrtl.fr/corpus/frantext/
Google Books	Varied writing	45,000,000,000	Online	https://books.google.com/ngrams
Lextutor	Varied writing	N.R.	Online	http://lextutor.ca/conc/fr/
Le Monde	Journalistic writing	500,000,000	Purchase only	ELRA
Est Républicain	Journalistic writing	N.R.	Downloadable	Ortlolang
Sciences Humaines	Journalistic writing	170,000	Downloadable	Ortolang
CoMeRe	New media writing	74,000,000	Downloadable	Ortolang
Backbone	Spoken regional varieties	N.R.	Downloadable	http://webapps.ael.uni-tuebingen.de/backbone-search/faces/search.jsp
CLAPI	Spoken French	46 hours	Online	http://clapi.ish-lyon.cnrs.fr/
CFPQ	Spoken Quebec French	700,000	Online	https://recherche.flsh.usherbrooke.ca/cfpq/
CFPP2000	Spoken Parisian French	700,000	Online	http://cfpp2000.univ-paris3.fr/
OFROM	Spoken Swiss French	1,000,000	Online	http://www11.unine.ch/index.php
VALIBEL	Spoken Belgian French	4,000,000	On demand	https://uclouvain.be/fr/instituts-recherche/ilc/valibel/
PFC	Mixed	1,800,000	Downloadable	Ortolang
CHILDES	Children's	Varied corpus	Downloadable or online	http://childes.psy.cmu.edu/browser/
EMA écrits scolaires	Children's	N.R.	Downloadable	Ortolang
Littéracie avancée	Young adults	900 000	Downloadable	Ortolang
Lund CEFLE	Swedish students	100,000	Partially downloadable	http://projekt.ht.lu.se/cefle/

Dire Autrement	English students	50,000	On demand	http://web5.uottawa.ca/direautrement/index.html
Interfra	Students	Varied corpus	Downloadable or online	https://www.su.se/romklass/interfra
French Learner Language Corpora	Students	Varied corpus	Downloadable or online	http://www.flloc.soton.ac.uk/
UWI French L2 Corpus	English and Creole students	15,000	Downloadable or online	https://slabank.talkbank.org/access/French/UWI.html
Cabal2	French-English parallel	400,000	Online	http://cabal.rezo.net/
Europarl	Multilingual parallel	–	Downloadable	http://www.statmt.org/europarl/
Hansard	French-English parallel	–	Downloadable	http://www.isi.edu/natural-language/download/hansard/
JRC-Acquis	Multilingual	–	Downloadable	https://ec.europa.eu/jrc/en/language-technologies/jrc-acquis
Ted Talks	Multilingual	–	Online	https://yohasebe.com/tcse/

Note.– In the table, the "–" symbol shows that a precise size cannot be indicated because the number of words is constantly evolving.

Chapter 6

Tools for building a corpus

AntFileConverter: http://www.laurenceanthony.net/software/antfileconverter/.

Sketch Engine: https://www.sketchengine.eu/.

WordSmith: https://www.lexically.net/wordsmith/.

Chapter 7

Standardization projects

Comité Language resource management: https://www.iso.org/committee/297592.html.

Projet Universal Dependencies: https://www.iso.org/committee/297592.html.

Tools for annotating a corpus

Brat: https://brat.nlplab.org/.

Dublin Core: http://dublincore.org/.

EXMARaLDA: https://exmaralda.org/en/.

GATE: https://gate.ac.uk/.

NITE XML Toolkit: http://groups.inf.ed.ac.uk/nxt/.

NLTK: https://www.nltk.org/.

SpaCy: https://spacy.io/.

Tool for the statistical computation of the agreement between annotators

VassarStats: http://vassarstats.net/.

Chapter 8

Tools for carrying out statistical tests online

GraphPad: https://www.graphpad.com/quickcalcs/.

Lancaster stats tool online: http://corpora.lancs.ac.uk/stats/toolbox.php.

VassarStats: http://vassarstats.net/.

Software for carrying out statistical tests

R (free software): https://cran.r-project.org/.

SPSS (fee-paying software): https://www.ibm.com/products/spss-statistics.

Index

D, E, F

G, H, I

K, L, M

N, O, P

Q, R, S

T, V, W

X, Y

Other titles from

ISTE

in

Cognitive Science and Knowledge Management

2020

OSIURAK François
The Tool Instinct

2019

CLAVEL Chloé
Opinion Analysis in Interactions: From Data Mining to Human-Agent Interaction

KAROUI Jihen, BENAMARA Farah, MORICEAU Véronique
Automatic Detection of Irony: Opinion Mining in Microblogs and Social Media

MARTINOT Claire, BOŠNJAK BOTICA Tomislava, GEROLIMICH Sonia, PAPROCKA-PIOTROWSKA Urszula
Reformulation and Acquisition of Linguistic Complexity: Crosslinguistic Perspective
(Interaction of Syntax and Semantics in Discourse Set – Volume 2)

2018

BONFANTE Guillaume, GUILLAUME Bruno, PERRIER Guy
Application of Graph Rewriting to Natural Language Processing
(Logic, Linguistics and Computer Science Set – Volume 1)

PENNEC Blandine
Discourse Readjustment(s) in Contemporary English
(Interaction of Syntax and Semantics in Discourse Set – Volume 1)

2017

KURDI Mohamed Zakaria
Natural Language Processing and Computational Linguistics 2: Semantics, Discourse and Applications

MAESSCHALCK Marc
Reflexive Governance for Research and Innovative Knowledge
(Responsible Research and Innovation Set – Volume 6)

PELLÉ Sophie
Business, Innovation and Responsibility
(Responsible Research and Innovation Set - Volume 7)

2016

BOUVARD Patricia, SUZANNE Hervé
Collective Intelligence Development in Business

CLERC Maureen, BOUGRAIN Laurent, LOTTE Fabien
Brain–Computer Interfaces 1: Foundations and Methods
Brain–Computer Interfaces 2: Technology and Applications

FORT Karën
Collaborative Annotation for Reliable Natural Language Processing

GIANNI Robert
Responsibility and Freedom
(Responsible Research and Innovation Set – Volume 2)

GRUNWALD Armin
The Hermeneutic Side of Responsible Research and Innovation (Responsible Research and Innovation Set – Volume 5)

KURDI Mohamed Zakaria
Natural Language Processing and Computational Linguistics 1: Speech, Morphology and Syntax

LENOIR Virgil Cristian
Ethical Efficiency: Responsibility and Contingency (Responsible Research and Innovation Set – Volume 1)

MATTA Nada, ATIFI Hassan, DUCELLIER Guillaume
Daily Knowledge Valuation in Organizations

NOUVEL Damien, EHRMANN Maud, ROSSET Sophie
Named Entities for Computational Linguistics

PELLÉ Sophie, REBER Bernard
From Ethical Review to Responsible Research and Innovation (Responsible Research and Innovation Set - Volume 3)

REBER Bernard
Precautionary Principle, Pluralism and Deliberation (Responsible Research and Innovation Set – Volume 4)

SILBERZTEIN Max
Formalizing Natural Languages: The NooJ Approach

2015

LAFOURCADE Mathieu, JOUBERT Alain, LE BRUN Nathalie
Games with a Purpose (GWAPs)

SAAD Inès, ROSENTHAL-SABROUX Camille, GARGOURI Faïez
Information Systems for Knowledge Management

2014

DELPECH Estelle Maryline
Comparable Corpora and Computer-assisted Translation

FARINAS DEL CERRO Luis, INOUE Katsumi
Logical Modeling of Biological Systems

MACHADO Carolina, DAVIM J. Paulo
Transfer and Management of Knowledge

TORRES-MORENO Juan-Manuel
Automatic Text Summarization

2013

TURENNE Nicolas
Knowledge Needs and Information Extraction: Towards an Artificial Consciousness

ZARATÉ Pascale
Tools for Collaborative Decision-Making

2011

DAVID Amos
Competitive Intelligence and Decision Problems

LÉVY Pierre
The Semantic Sphere: Computation, Cognition and Information Economy

LIGOZAT Gérard
Qualitative Spatial and Temporal Reasoning

PELACHAUD Catherine
Emotion-oriented Systems

QUONIAM Luc
Competitive Intelligence 2.0: Organization, Innovation and Territory

2010

ALBALATE Amparo, MINKER Wolfgang
Semi-Supervised and Unsupervised Machine Learning: Novel Strategies

BROSSAUD Claire, REBER Bernard
Digital Cognitive Technologies

2009

BOUYSSOU Denis, DUBOIS Didier, PIRLOT Marc, PRADE Henri
Decision-making Process

MARCHAL Alain
From Speech Physiology to Linguistic Phonetics

PRALET Cédric, SCHIEX Thomas, VERFAILLIE Gérard
Sequential Decision-Making Problems / Representation and Solution

SZÜCS Andras, TAIT Alan, VIDAL Martine, BERNATH Ulrich
Distance and E-learning in Transition

2008

MARIANI Joseph
Spoken Language Processing